Urban Sustainability

The Urban Sustainability Book Series is a valuable resource for sustainability and urban-related education and research. It offers an inter-disciplinary platform covering all four areas of practice, policy, education, research, and their nexus. The publications in this series are related to critical areas of sustainability, urban studies, planning, and urban geography.

This book series aims to put together cutting-edge research findings linked to the overarching field of urban sustainability. The scope and nature of the topic are broad and interdisciplinary and bring together various associated disciplines from sustainable development, environmental sciences, urbanism, etc. With many advanced research findings in the field, there is a need to put together various discussions and contributions on specific sustainability fields, covering a good range of topics on sustainable development, sustainable urbanism, and urban sustainability. Despite the broad range of issues, we note the importance of practical and policy-oriented directions, extending the literature and directions and pathways towards achieving urban sustainability.

The series will appeal to urbanists, geographers, planners, engineers, architects, governmental authorities, policymakers, researchers of all levels, and to all of those interested in a wide-ranging overview of urban sustainability and its associated fields. The series includes monographs and edited volumes, covering a range of topics under the urban sustainability topic, which can also be used for teaching materials.

Ali Cheshmehzangi • Sara Alidoust
Wendy Y. Chen • Richard Fuller
Editors

Designing Nature into Cities

Springer

Editors
Ali Cheshmehzangi
School of Architecture, Design and Planning
University of Queensland
Brisbane, QLD, Australia

Sara Alidoust
School of Architecture, Design and Planning
University of Queensland
Brisbane, QLD, Australia

Wendy Y. Chen
Department of Geography
University of Hong Kong
Hong Kong, Hong Kong

Richard Fuller
School of the Environment
University of Queensland
Brisbane, QLD, Australia

ISSN 2731-6483 ISSN 2731-6491 (electronic)
Urban Sustainability
ISBN 978-981-95-9381-1 ISBN 978-981-95-9382-8 (eBook)
https://doi.org/10.1007/978-981-95-9382-8

This Springer imprint is published by the registered company Springer Nature Singapore Pte Ltd.
The registered company address is: 152 Beach Road, #21-01/04 Gateway East, Singapore 189721, Singapore

To those who plant, protect, and persist.
To every hand that restores a wetland,
designs a green roof, or reclaims a forgotten
stream, this book is for you.

Preface

> Be like the sun for grace and mercy. Be like the night to cover others' faults. Be like running water for generosity. Be like death for rage and anger. Be like the Earth for modesty. Appear as you are. Be as you appear.
>
> —Rumi, the well-known Persian Poet

We stand at a decisive threshold in the story of cities. For generations, urban progress has been measured in towers raised, roads extended, and systems accelerated. Steel and concrete have been our monuments; efficiency and scale our guiding myths. Yet in the pursuit of mastery over space, we quietly severed our bond with the living systems that made cities possible in the first place. In many places around the globe, nature was reduced to scenery, an accessory at the edge of development rather than the foundation beneath it. The result is now impossible to ignore: heat trapped between walls, water rushing where it once rested, air thickened by excess, and communities disconnected from the rhythms that once gave them coherence and care.

Designing Nature into Cities is written for this moment of reckoning. It is a statement that cities need to be rebuilt on a deeper logic, one that views ecological processes as the foundation of urban form rather than its ornamentation, rather than a call to abandon urban existence. This book maintains that nature is civilization's oldest and most trustworthy engineer rather than its backdrop. Rivers are already adept at handling water. Forests are already adept at cooling the air. Wetlands are already adept at absorbing shock. These days, design needs to learn how to communicate, listen, and work together.

Fundamentally, this book makes the straightforward but daring assertion that using nature as a design guide is essential to creating resilient and liveable communities in the future. Instead of being heat traps, streets can become climate buffers. Instead of becoming passive surfaces, roofs may develop into pollinator pathways. Rather than being buried beneath it, rivers can resurface as civic areas that shape public life. Every one of these changes redefines infrastructure as a living system as opposed to a mechanical one, with each intervention supporting both ecological continuity and human needs.

The book explores case studies from throughout the world where the distinction between natural process and built form is purposefully blurred. Instead of being an ecological expense, design becomes a tool for ecological repair in these settings. Neighbourhoods are cooled by plant intelligence rather than just equipment. Instead of being drained for convenience, districts are redesigned around regenerated watersheds. These gestures are not symbolic. Because they collaborate with nature rather than oppose it, they are functional models of cities that perform better.

Technology also appears in this narrative, but it does so as an interpreter of ecology rather than as a replacement. Designers can now anticipate the long-term ecological effects of today's decisions thanks to new digital tools, which show how decades of water, heat, and biodiversity might be affected by a single material choice. Speculation is replaced with forecasting. Isolated fixes are replaced with systems thinking. Instead of being an act of spontaneity, design becomes an act of responsibility. Yet this book reaches beyond technique. It addresses governance and culture as much as form. Ecological cities are not built by drawings alone. They require policies that protect long-term value over short-term gain, and communities that see themselves as stewards rather than consumers of place. When residents plant, manage, and protect shared landscapes, cities gain more than green cover. They gain continuity, trust, and identity. Nature becomes not only infrastructure, but also a narrative; a shared language through which people understand where they belong.

In this vision, every building is a potential habitat. Every road is a chance for cooling and filtration. Every public space is an opportunity for human and non-human life to meet without conflict. Design ceases to be an exercise in domination and becomes an act of choreography, aligning human movement with ecological flow. This is not a nostalgic return to some imagined past, but a disciplined step towards an intelligent future.

The urgency of this transformation cannot be overstated. Climate volatility, demographic pressure, and resource exhaustion are rewriting the terms of urban survival. Cities will either adapt by aligning with natural systems or struggle beneath the weight of their own rigidity. The choice is no longer aesthetic. It is existential. Nature is not an external force pressing upon cities from outside. It is already inside them, waiting to be acknowledged as a co-author of urban life.

Designing Nature into Cities completes a trilogy that moves from recognition, to integration, to embodiment. Here, nature is no longer invited in as a guest. It is installed as a structural partner. The ambition is not simply greener cities, but wiser ones, places where growth does not erase memory, where innovation does not require sacrifice, and where resilience is measured in living capacity rather than mechanical output.

This book exhorts us to have the guts to reconsider the purpose of cities, to design with humility, and to have faith in systems that predate modern technologies. By doing this, it presents a future where urban life is ingrained in nature rather than suspended above it. It is not an unrealistic notion for a city to breathe, filter, shelter, and renew alongside its inhabitants. It is the most practical reaction to the upcoming century.

The task before us is not to build more, but to build more wisely. To design not only for speed and scale, but for grace and mercy, generosity, and modesty; qualities that Rumi saw as the measure of a balanced life. In cities shaped by these principles, nature does not compete with humanity. It completes it.

Brisbane, Australia — Ali Cheshmehzangi
Brisbane, Australia — Sara Alidoust
Hong Kong, Hong Kong — Wendy Y. Chen
Brisbane, Australia — Richard Fuller

Acknowledgements

We acknowledge everyone's contributions to this exciting book project. This book is the last of three volumes we put together on the topic of 'Nature in Cities'. We extend our gratitude to all the authors, co-authors, chapter contributors, and collaborators who have shared their expertise, passion, and creativity with us. This sentiment applies to our earlier and forthcoming work as well, where we deeply value the dedication and vision of all contributors who helped create this book. Such collaborative efforts foster new dialogues, networking opportunities, and a valuable research exchange platform for scholars worldwide.

About the Book

As the final volume in this trilogy, *Designing Nature into Cities* advances the conversation from integrating nature into urbanism to embedding it as a primary design principle. This book argues that the future of resilient and liveable cities depends on making ecological systems not a supplement to design, but the scaffolding upon which urban life is built. The book explores how communities, engineers, planners, and architects can collaborate to regard streets as climate buffers, rooftops as pollinator habitats, and rivers as civic spaces, where each design choice benefits the ecosystem.

The book explores global case studies where the line between the built and natural environment is deliberately blurred. These examples demonstrate how design can be a tool for ecological recovery, from neighbourhoods that use natural plants as a passive cooling system to cities that have completely remade districts around restored watersheds. It explores the technology tools that are allowing designers to forecast long-term ecological implications prior to construction.

Beyond infrastructure, the volume explores the governance and cultural frameworks that enable these changes, demonstrating how community stewardship and innovative policymaking can lock in ecological gains over many years. It views nature as an active partner in forming urban futures rather than just an aesthetic effect or source of environmental services. The book presents a vision of urban settings in which every building, roadway, and public area is transformed into a location where nature can coexist peacefully with human activity.

Key features and benefits:

- *Showcases innovative methods of designing nature into cities and communities*
- *Designing cities where nature is a core blueprint, not an afterthought.*
- *Offers practical, scalable strategies for embedding nature at every city scale.*

Contents

Heritage, Identity and Urban Transformation

Edge Ecologies: Coasts and Deserts

Editors and Contributors

About the Editors

Ali Cheshmehzangi is Head of the School of Architecture, Design and Planning and a Professor at University of Queensland (UQ), Brisbane, Australia. He is among the top 20 global scholars in the urban sustainability research area. He is the Editor-in-Chief of the Urban Sustainability Book Series with Springer Nature. With a career spanning over two decades, he has made significant contributions to the academic and professional communities, with a focus on sustainable and environmentally conscious design. So far, Ali has published over 650 journal papers, articles, conference papers, book chapters, and reports. He has published more than 58 academic books, some of which have received awards at the international, national, provincial, and municipal levels. He has also received international awards and recognition for his research on urban resilience studies and sustainability research, as well as the 2018 Vice-Chancellor's award for his impactful contribution to higher education.

Sara Alidoust is a Senior Lecturer in Urban Planning at School of Architecture, Design and Planning at The University of Queensland (UQ), Brisbane, Australia. Her research centres on the intersections of Urban Planning and Public Health, with a strong emphasis on interdisciplinary collaboration and practical application. Sara delivers transdisciplinary solutions to some of the most compelling challenges of our time, 'housing equity' and 'healthy cities'. Her work stands out through the application of Systems Thinking to complex planning issues and she explores the connections between planning and the physical, mental, and social health of individuals and communities. Her primary focus is on developing resilient cities, capable of maintaining their liveability over time, especially amidst disruptive events and shocks.

Wendy Y. Chen is a professor in the Department of Geography at University of Hong Kong. She serves as the Editor-in-Chief for Urban Forestry & Urban Greening, a top international journal in the field of urban greening design and management. Her research agenda has focused on key knowledge gaps in evaluating and modelling urban green-blue spaces (GBS) that are reserved, modified, and deployed in urban/peri-urban areas. She is amongst a small group of scholars who have begun to experiment and validate the utilization of classical non-market approaches in China's transitional context and facilitate the quantitative assessment of GBS's contribution to the quality of life. An additional strand of her research is dedicated to investigating GBS dynamics, pertaining to how and why GBS as public environmental goods are physically transformed, economically incorporated, and socially mobilized. Her research helps to integrate socioeconomic dimensions to enrich GBS scholarship, and rethink theories pertaining to GBS demand and supply in the context of extensive urbanization in the developing countries and re-urbanization in the developed world. She is ranked as a top-1% scholar worldwide by Clarivate Analytics.

Richard Fuller is a leading expert in urban ecology research. He is a Professor at School of the Environment at University of Queensland (UQ), Brisbane, Australia. He works on pure and applied topics in biodiversity and conservation. Much of his work is interdisciplinary, focusing on the interactions between people and nature, how these can be enhanced, and how these relationships can be shaped to converge on coherent solutions to the biodiversity crisis. His current research topics include the ecology of urban environments, understanding what drives some people to show stronger environmental concern than others, and strategies for designing efficient conservation plans. He enjoys working closely with all his wonderful colleagues in the Centre for Biodiversity and Conservation Science.

Contributors

Sara Alidoust School of Architecture, Design and Planning, University of Queensland, Brisbane, QLD, Australia

Hooshmand Alizadeh Faculty of Art and Architecture, Department of Urban Planning and Design, University of Kurdistan, Sanandaj, Iran

Department of Geography and Regional Research, University of Vienna, Vienna, Austria

Necdet Ayik Department of Architecture and Urban Studies, Politecnico di Milano, Milan, Italy

Wendy Y. Chen Department of Geography, University of Hong Kong, Hong Kong, Hong Kong

Ali Cheshmehzangi School of Architecture, Design and Planning, University of Queensland, Brisbane, QLD, Australia

School of Architecture and Urban Planning, Hunan University, Changsha, China

College of Architecture and Urban Planning (CAUP), Tongji University, Shanghai, China

Milad Doostvandi Faculty of Art and Architecture, Department of Urban Planning and Design, University of Kurdistan, Sanandaj, Iran

Junchi Fang College of Agriculture and Life Sciences, Cornell University, Ithaca, NY, USA

Tasnim Ferdous University of Manitoba, Winnipeg, MB, Canada

Richard Fuller School of the Environment, University of Queensland, Brisbane, QLD, Australia

Centre for Biodiversity and Conservation Science, The University of Queensland, Brisbane, Australia

Jane Handal Department of Architecture, Prince Sultan University, Riyadh, Saudi Arabia

Huixin Huang College of Environmental Design, University of California, Berkeley, CA, USA

Erna Husukic Department of Architecture, Prince Sultan University, Riyadh, Saudi Arabia

Muhammad Nelza Mulki Iqbal School of Architecture and Landscape University of Sheffield, Sheffield, UK

Institut Teknologi Nasional Malang, Malang, Indonesia

Lu Jin School of Architecture, Design and Planning, University of Queensland, Brisbane, Australia

Zachary M. Jones Department of Architecture and Urban Studies, Politecnico di Milano, Milan, Italy

Abdallah Jreij Department of Architecture and Urban Studies, Politecnico di Milano, Milan, Italy

Israa Mahmoud Laboratorio di Simulazione Urbana (Fausto Curti), Department of Architecture and Urban Studies, Politecnico di Milano, Milan, Italy

Silvia Micheli School of Architecture, Design and Planning, University of Queensland, Brisbane, Australia

Monisha Momtaz Department of Architecture, Bangladesh University, Dhaka, Bangladesh

Monica Moscatelli Department of Architecture, Prince Sultan University, Riyadh, Saudi Arabia

Ayda Rastiemadabadi Department of Urbanism, Territory and Landscape, Polytechnic University of Catalonia, Barcelona, Spain

Susanne Schmidt School of Agriculture and Food Sustainability, University of Queensland, Brisbane, Australia

Jamshid Sodikov Tashkent State Transport University, Tashkent, Uzbekistan

Manqi Yang Faculty of Social and Behavioral Science, Universiteit van Amsterdam, Amsterdam, Netherlands

Milad Zandsalimi Faculty of Art and Architecture, Department of Urban Planning and Design, University of Kurdistan, Sanandaj, Iran

Nature as Design: Recasting the Urban Blueprint

Ali Cheshmehzangi, Sara Alidoust, Wendy Y. Chen, and Richard Fuller

Abstract This book is the final volume in the *Nature and Cities* trilogy, advancing the narrative from nurturing and integrating nature into urbanism to embedding it as the blueprint of city-making. This book argues that the future of resilient and liveable cities depends on treating ecological systems not as supplemental features but as the scaffolding upon which urban life is designed and sustained.

Through global case studies, the volume explores how architecture, infrastructure, and public spaces can be reconceived as living ecologies: streets functioning as climate buffers, rooftops as pollinator habitats, and rivers as civic spaces. It highlights how communities, planners, architects, and policymakers can collaborate to create cities that regenerate ecosystems, strengthen resilience, and foster inclusion. By embracing design as dialogue with nature, this volume positions ecological principles as central to long-term climate adaptability and cultural identity.

As the culmination of the trilogy, *Designing Nature into Cities* presents both a vision and a roadmap for regenerative urbanism. It situates itself within global sus-

A. Cheshmehzangi (✉)
School of Architecture, Design and Planning, University of Queensland, Brisbane, QLD, Australia

School of Architecture and Urban Planning, Hunan University, Changsha, China

College of Architecture and Urban Planning (CAUP), Tongji University, Shanghai, China
e-mail: a.chesh@uq.edu.au

S. Alidoust
School of Architecture, Design and Planning, University of Queensland, Brisbane, QLD, Australia
e-mail: s.alidoust@uq.edu.au

W. Y. Chen
Department of Geography, University of Hong Kong, Hong Kong, Hong Kong

R. Fuller
School of the Environment, University of Queensland, Brisbane, QLD, Australia

Centre for Biodiversity and Conservation Science, The University of Queensland, Brisbane, Australia

A. Cheshmehzangi et al. (eds.), *Designing Nature into Cities*, Urban Sustainability, https://doi.org/10.1007/978-981-95-9382-8_1

tainability discourse as a call to rethink urban futures: not as machines for growth, but as living, adaptive ecologies where human and non-human life flourish together.

Keywords Nature-based design · Ecological blueprint · Regenerative urbanism · Climate resilience · Urban liveability · Biodiversity · Public space · Nature-based solutions · NBS

1 Introducing Nature as Design: The Next Frontier

Cities have always been designed. From the gridded streets of ancient Mesopotamia to the master-planned capitals of the twentieth century, design has been the language through which societies imagine their futures. Yet in most traditions of city-making, design has privileged geometry, efficiency, and order over ecological relationships. Nature has been treated as background, a source of inspiration or amenity, but rarely as a central actor in shaping the blueprint of urban form. This volume, the third in the *Nature and Cities* trilogy, argues for a decisive shift: from integrating nature into cities as an added layer, to designing cities where nature is the foundation upon which urban life is built.

The first book, *Nature in Cities, Nurturing Cities*, established the philosophical and ethical ground. It asked what it means to nurture cities through ecological care, positioning nature as vital to resilience, wellbeing, and justice. The second, *Nature-based Urbanism*, demonstrated how that vision translates into practice, showcasing global examples of ecological restoration, biophilic design, and nature-based solutions embedded into planning and governance. This final volume advances the conversation further. *Designing Nature into Cities* takes the step from integration to blueprint, from supplement to generative principle. Its central claim is bold: the future of resilient and liveable cities depends on designing with nature as the guiding framework, not as a marginal adjustment.

This is not a rhetorical flourish but a recognition of the challenges of the twenty-first century. Climate change is intensifying heat waves, floods, and storms. Biodiversity loss is eroding ecological resilience, while urban sprawl and infrastructure expansion continue to fragment habitats (Johnson & Klemens, 2005; Simkin et al., 2022; Ren et al., 2023). Public health is threatened not only by pollution but by the mental and physical costs of disconnection from living systems. Conventional urban design, dominated by steel, concrete, and asphalt, cannot meet these challenges alone. What is required is an ecological blueprint, where every design decision is evaluated by its capacity to sustain and regenerate the natural systems upon which cities depend.

Designing with nature means treating streets as climate buffers, not just traffic corridors. It means imagining rooftops as habitats for pollinators, not as wasted space above air-conditioning units. It requires rivers to be restored as civic spaces that carry memory, culture, and biodiversity rather than confined to channels of drainage. Buildings become adaptive organisms, shaded and cooled by plants, while public squares double as sponges for rainfall and sanctuaries for social interaction

(Chan et al., 2022; Li et al., 2023, 2026). In this paradigm, the built environment is no longer set against nature but co-created with it.

Crucially, this approach is not limited to design aesthetics. It redefines the very process of city-making. Engineers, planners, architects, and communities must work collaboratively with ecologists, climatologists, and cultural custodians to ensure that design choices contribute to both ecological and human flourishing. Digital technologies, from ecological modelling to AI-driven scenario planning, extend the designer's capacity to forecast long-term environmental impacts, embedding foresight into every blueprint. Governance frameworks must follow, ensuring that design interventions are supported by policies that sustain ecological gains over generations.

The urgency of embedding nature into design is matched by its transformative potential. Global case studies already point the way. Districts rebuilt around restored watersheds demonstrate how design can anchor resilience. Neighbourhoods using native vegetation as passive cooling systems reduce energy demand while enhancing cultural identity (Anandani, 2024). Entire cities, from Singapore's biophilic networks (Wong, 2024) to Medellín's green corridors (Paniagua-Villada et al., 2024), reveal that designing with nature can transform not only physical form but civic imagination. These examples show that the line between the natural and the built is not fixed but can be deliberately blurred, generating places where ecological health and human vitality reinforce one another.

As the culmination of this trilogy (see Cheshmehzangi et al., 2026a, 2026b), *Designing Nature into Cities* positions itself within the global discourse on climate resilience and urban liveability as both visionary and practical. It insists that ecological systems are not side conditions to be managed but the scaffolding for sustainable futures. By embedding nature as a primary design principle, this book challenges architects, planners, policymakers, and communities to reconsider their roles. They are not only designers of structures but stewards of ecosystems, co-authors of a shared urban ecology.

The next frontier of city-making is not technological spectacle or endless expansion but the ability to design for regeneration. This is the horizon this book sets: cities conceived not as machines for growth, but as living ecologies that breathe, adapt, and endure. In this vision, design is not merely about shaping space; it is about sustaining life.

1.1 From Integration to Blueprint: Strengthening the Narrative

The language of "integration" has dominated recent conversations about nature in cities (Cook & Swyngedouw, 2012; Simeonova & Van Der Valk, 2016; Gentin et al., 2019; Grace et al., 2023). From green roofs to stormwater gardens, cities have learned to weave ecological elements into their infrastructure. This has been a vital shift, breaking with the industrial era when nature was sidelined or erased from urban form. Yet integration alone is not enough. Too often, it has produced scattered interventions, e.g. parks inserted into dense grids, rain gardens tucked between asphalt expanses, or decorative plantings masking unsustainable systems. These gestures soften the city but rarely transform it. The challenge now is to move from integration to blueprint: to make nature the primary framework through which design, planning, and governance unfold.

This distinction is more than semantic. Integration suggests addition, i.e. nature layered onto existing systems. A blueprint implies origin, i.e. nature shaping the very logic of those systems. When ecological flows, biodiversity, and resilience become the starting point of design, the city is reimagined from the ground up. Streets are not widened and then "greened"; they are conceived as ecological corridors. Housing is not built and later retrofitted with plants; it is designed with natural ventilation, shading, and habitats in mind. Water is not drained away through pipes and channels; it is absorbed, filtered, and celebrated as part of civic life. In blueprint thinking, nature is not an accessory but the structure itself.

This narrative shift matters because the limits of integration are increasingly visible. Token greening has not prevented the intensification of urban heat islands. Isolated parks have not halted biodiversity decline. Grey infrastructure reinforced with green accents still struggles to manage extreme floods and rising seas. The twenty-first century demands systemic solutions, not cosmetic ones. Blueprint design, by embedding ecological principles into every layer of city-making, offers that systemic shift.

Global examples already hint at this trajectory. In Copenhagen, cloudburst boulevards are designed as dual-purpose streets and flood channels, reconfiguring mobility and water management together. In China's Sponge City programme, water-sensitive design is no longer an add-on but a mandatory baseline (Sun et al., 2022). Singapore's network of connected parks and corridors exemplifies blueprint thinking, where the city's structure follows ecological continuity rather than imposing fragmentation. These examples demonstrate that nature-based design can be codified into policy, scaled into practice (Li et al., 2021), and normalised as a cultural expectation.

Moving from integration to blueprint also transforms professional roles. Architects and planners become not only designers of form but custodians of ecosystems. Engineers calculate not just load-bearing structures but ecological performance. Policymakers are asked not merely to regulate land use but to steward long-term resilience. Communities, too, are no longer passive recipients of greening projects but co-creators of local ecologies, shaping and caring for landscapes that sustain their wellbeing.

The ambition of this book is to strengthen this narrative—to show that designing nature into cities is not a utopian dream but a practical, scalable, and necessary response to converging global crises. It is to insist that the next era of urbanism will be defined not by integration at the margins but by ecological blueprints at the core.

This volume thus builds on the first two books of the trilogy. *Nurturing Cities* asked us to care; *Nature-based Urbanism* showed us pathways; *Designing Nature*

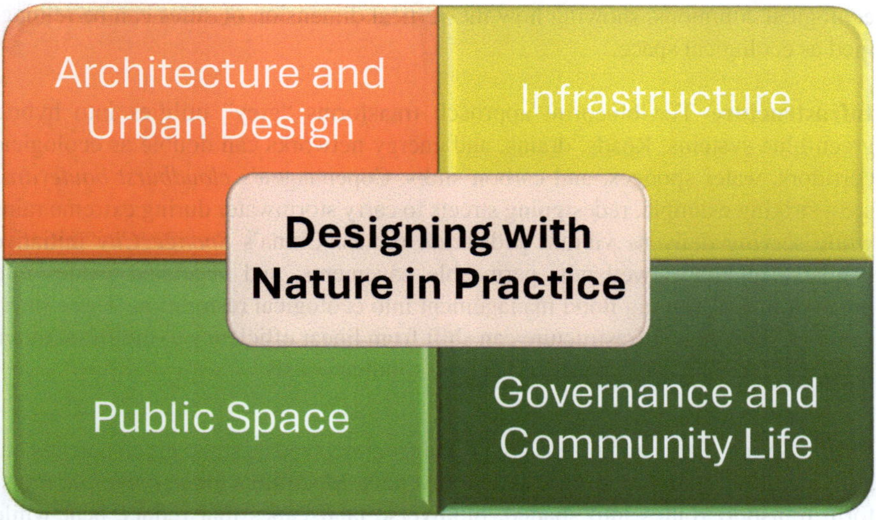

Fig. 1 Summary of four key areas when designing with nature in practice

into Cities now provides the blueprint. The arc is deliberate: from philosophy, to practice, to design principle. Together, they offer not only a vision of what cities could become but a roadmap for how to get there.

2 Designing with Nature in Practice: Transforming Architecture, Infrastructure, and Public Space

Designing nature into cities becomes real when ideas leave theory and take form in the built environment. This paradigm asks not for gestures of greening but for the transformation of architecture, infrastructure, and public space into systems that operate as living ecologies. The goal is not merely to soften the hard edges of urban life but to design environments where human and non-human life can coexist, regenerate, and thrive (See Fig. 1).

Architecture and Urban Design Buildings designed with nature at their core blur the distinction between interior and exterior, structure and habitat. Green façades and vertical forests, such as Milan's *Bosco Verticale*, demonstrate how dense housing can host biodiversity while improving air quality and moderating temperature. Passive cooling systems that use vegetation, shading, and natural ventilation reduce energy demand while enhancing comfort, a principle visible in vernacular traditions and now re-emerging in contemporary bioclimatic design. Rooftops and terraces increasingly serve as pollinator habitats, food-producing gardens, or communal ecological commons, showing how the vertical dimension of cities can be reimagined as ecological space.

Infrastructure The blueprint approach transforms "grey" utilities into hybrid green-blue systems. Roads, drains, and energy networks can double as ecological corridors, water sponges, and carbon sinks. Copenhagen's *cloudburst boulevards* are a striking example, redesigning streets to carry stormwater during extreme rains while serving daily as vibrant pedestrian routes. China's *Sponge City* initiative similarly integrates wetlands, permeable pavements, and vegetated swales into urban drainage, turning flood management into ecological restoration. These interventions show how infrastructure can shift from linear efficiency to multifunctional resilience, serving ecosystems and people simultaneously.

Public Space Parks and plazas are no longer conceived as static green patches but as dynamic ecologies of culture and biodiversity. Medellín's *green corridors* transform transport routes into shaded, biodiverse landscapes that reduce heat while connecting communities. Singapore's *Gardens by the Bay* illustrates how large-scale ecological design can create iconic civic identity while performing ecological functions. Smaller interventions, like community-led gardens in Latin American neighbourhoods, reveal how public space designed with nature fosters stewardship, inclusion, and belonging. Public spaces become not only sites of recreation but platforms of ecological justice, ensuring that access to nature is distributed equitably.

Governance and Community Life Designing with nature in practice requires governance frameworks that lock ecological gains into long-term strategies. This means building codes that demand ecological performance, planning regulations that prioritise connectivity of green networks, and funding mechanisms that value ecosystem services. Equally, it requires cultural shifts: communities participating in co-design and stewardship, reclaiming agency in shaping their neighbourhood ecologies. In Tokyo, citizen groups actively manage pocket wetlands; in Melbourne, participatory planning has shaped extensive green-blue infrastructure projects. These examples show that ecological design endures only when it becomes part of shared civic culture.

Taken together, these practices point to a profound shift in city-making. Architecture becomes habitat, infrastructure becomes ecosystem, and public space becomes a platform for justice and belonging. Designing with nature in practice is not about applying a universal model but about creating context-specific ecologies that are adaptive, resilient, and inclusive. It is through this lens that the book examines global case studies, showing how the ecological blueprint is already emerging—and how it can be scaled to define the future of urban design.

3 Purpose and Ambition of This Volume: The Ecological Blueprint

The purpose of this third and final volume in the *Nature and Cities* trilogy is to bring the conversation to its most decisive stage: to argue that nature is not merely to be nurtured or integrated, but to be embedded as the primary blueprint of city-making. While the first book, *Nature in Cities, Nurturing Cities*, framed the ethical imperative of ecological care, and the second, *Nature-based Urbanism*, demonstrated how integration is being put into practice, *Designing Nature into Cities* takes the decisive step of positioning design itself as the medium through which ecological futures are created. This is the ecological blueprint—the claim that every design decision, from the layout of a street to the orientation of a building, from the management of water to the programming of public space, must be guided by ecological principles.

The intellectual aim of this volume is to establish "designing with nature" as the defining paradigm of the twenty-first-century urbanism. Historically, design has been seen as a human act imposed upon land and resources, with nature relegated to backdrop or constraint. This book challenges that paradigm. It reimagines design as a dialogue with ecological systems, where cities are not built *on top of* nature but *with and through* it. The ambition is to articulate a design philosophy that treats ecological flows, i.e., water, air, soil, biodiversity, as generative elements of urban form rather than as problems to be engineered away.

At a practical level, the objective is to show how this ecological blueprint can be applied at multiple scales. For architects, it means designing buildings as living habitats, adaptive to climate and generous to biodiversity. For planners, it means structuring neighbourhoods and districts around ecological networks rather than isolating them into fragments. For engineers, it means reconfiguring infrastructure as multifunctional, resilient, and regenerative. For communities, it means engaging in co-design and stewardship, ensuring that ecological gains are not only technical but cultural, rooted in shared identity and collective responsibility.

The book also positions itself firmly within the global urban sustainability discourse. Much of that discourse has focused on technology-driven solutions, smart systems, or economic innovation. While valuable, these approaches often risk marginalising the ecological foundations of urban life. *Designing Nature into Cities* insists that sustainability cannot be achieved without embedding ecological care at the centre of design (Williams, 2007; Beatley, 2011; Farr, 2011). It highlights how the ecological blueprint offers solutions that are not only climate-resilient but also socially inclusive, culturally meaningful, and economically viable. By bridging disciplines and showcasing global case studies, the book demonstrates that ecological design is not a peripheral experiment but an emerging global norm.

As the final step in the trilogy, this volume carries a particular ambition: to offer closure while opening horizons. It concludes the arc begun with nurturing and continued through practice, but it also projects forward, towards speculative, adaptive, and regenerative futures. In this sense, the ecological blueprint is not only a design principle but a cultural shift, a reorientation of values that insists cities are part of living systems. The ambition is to leave readers with a conviction that designing nature into cities is not optional, not decorative, but the condition of urban survival and flourishing.

This is why this volume matters. It provides both philosophical depth and practical strategies, situating itself as a call to action for designers, policymakers, and citizens. More than a book about design, it is a manifesto for a new urban condition—one where nature is not simply present in cities but where cities themselves become ecological blueprints of a regenerative future.

3.1 Structure of the Book: From Principles to Futures

This book is structured as a progressive journey, moving from conceptual foundations to applied strategies and, finally, to speculative and future-oriented reflections on designing with nature in urban contexts. Its organisation reflects a deliberate logic: to first establish why nature must be repositioned as a central design principle, then demonstrate how this principle is translated into material practices and spatial strategies and ultimately explore how these approaches can inform long-term urban futures under conditions of climate uncertainty, cultural transformation, and ecological stress.

This chapter serves as the introduction that frames the intellectual and ethical stance of the book. It redefines nature not as a decorative element or technical add-on, but as a formative force in shaping urban form, social life, and environmental performance. This chapter establishes the theoretical lens through which subsequent chapters should be read, emphasising nature as design logic rather than design supplement. At the other end of the volume, chapter "Reimagining Urban Futures: Lessons for Designing Nature into Cities" functions as a synthesis and projection. It draws lessons from the preceding chapters to articulate broader implications for urban practice, governance, and future design thinking, translating empirical insights into a forward-looking vision for nature-integrated cities.

Between these two anchors, the book is organised into three thematic parts, each comprising three case study-driven chapters (See Fig. 2). Together, they demonstrate how shared principles can generate diverse spatial outcomes across climatic, cultural, and socio-political contexts.

Part I, *Design Methods and Material Practices*, focuses on how nature-informed thinking reshapes the tools and techniques of urban design. The chapters in this section examine vegetation, materials, and analytical methods as active agents in placemaking. From subtropical cities shaped by native plant ecologies to contested material practices in Indonesian markets, and GIS-based approaches to biophilic streetscapes in Central Asia, this part shows how theory becomes operational through design methods. The emphasis is on experimentation, technical adaptation, and the translation of ecological knowledge into spatial form.

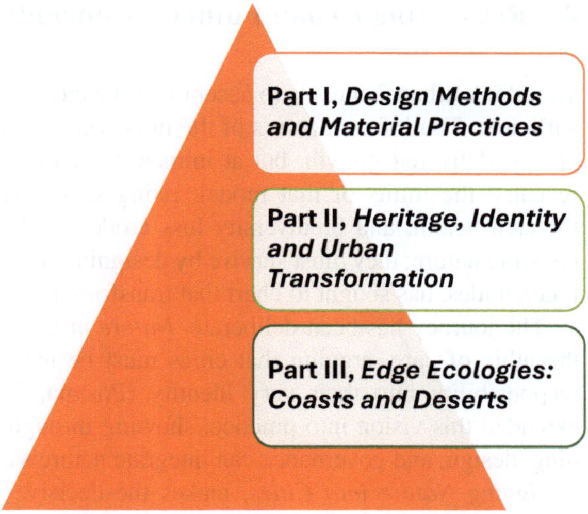

Fig. 2 The overall summary of three thematic parts of this volume

Part II, *Heritage, Identity, and Urban Transformation*, shifts attention to the cultural and historical dimensions of nature-driven placemaking. These chapters explore how natural systems intersect with memory, identity, and social meaning in cities undergoing change. Case studies from Genoa, Shiraz, and hilly urban environments demonstrate that nature is not only ecological infrastructure but also a cultural construct embedded in narratives of belonging, tradition, and modernisation. This part highlights the dialectical relationship between preservation and transformation, showing how nature can mediate between past urban identities and emerging spatial futures.

Part III, *Edge Ecologies: Coasts and Deserts*, addresses cities located in extreme or fragile environments, where ecological limits are most visible and design decisions carry heightened risk. Coastal and desert case studies from Bangladesh, Saudi Arabia, and other vulnerable contexts foreground nature-based solutions as instruments of resilience, adaptation, and social inclusion. Here, nature is presented not merely as a design strategy but as a survival framework, shaping how cities respond to sea-level rise, heat stress, and environmental degradation.

Taken together, the three parts form a coherent progression: from methods, to meanings, to margins. The case studies operate as both empirical evidence and design laboratories, illustrating how shared principles are reinterpreted across different geographies. This structure allows the book to bridge theory and practice, while also moving beyond documentation towards speculation. By culminating in a concluding chapter that synthesises insights across all parts, the book advances a vision in which designing with nature becomes a core urban paradigm, one capable of addressing present challenges while opening pathways towards more resilient, inclusive, and ecologically grounded urban futures.

4 Rethinking Urban Futures: Concluding Reflections

To rethink urban futures is to accept that the city, as we have known it, is no longer sufficient. The infrastructures of the industrial age, concrete, asphalt, grids of efficiency, delivered growth, but at immense ecological cost. The Anthropocene has revealed the limits of that model: rising seas overwhelm seawalls, heat islands threaten health, and biodiversity loss erodes resilience. Cities cannot survive by resisting nature; they must survive by designing with it. This volume, and the trilogy it concludes, has sought to chart that transformation.

The journey has been deliberate. *Nature in Cities, Nurturing Cities* established the ethic of care, arguing that cities must be nurtured by embedding ecological responsibility into their very identity (Postma, 2006). *Nature-based Urbanism* extended this vision into practice, showing through global case studies how planning, design, and governance can integrate nature as infrastructure. This final book, *Designing Nature into Cities*, makes the decisive claim: the blueprint of urban futures must be ecological. Nature is not an afterthought but the generative principle by which cities are designed, lived, and sustained.

This is more than a design philosophy; it is a reorientation of values. To design with nature is to view rivers as civic spaces, buildings as habitats, and public squares as ecologies of justice and belonging. It is to treat every street as a climate corridor and every roof as a living landscape. It is to imagine the city not as a machine but as an evolving organism, inseparable from the soil, air, water, and species that sustain it.

In the global discourse on climate resilience and liveability, this trilogy contributes a roadmap for regenerative urbanism. It insists that technological innovation and economic systems must be paired with ecological wisdom and cultural stewardship. It demonstrates that nurturing, integrating, and designing with nature are not sequential tasks but interwoven practices that together define the future of cities.

The ambition of this concluding reflection is both philosophical and practical. Philosophical, because it frames cities as moral projects, judged by how they sustain life. Practical, because it underscores the urgency: climate change, biodiversity loss, and inequality demand immediate and systemic action.

To rethink urban futures is to move beyond fear and limitation towards imagination and possibility. It is to accept that the most enduring cities will be those designed as living ecologies, i.e. adaptive, just, and regenerative. This trilogy offers no single template, but it does offer a compass: to build cities that are not only habitable, but profoundly alive.

References

Anandani, N. (2024). *Landscape design for passive cooling and energy conservation* (Master's thesis, University of Georgia), Georgia, USA.

Beatley, T. (2011). *Biophilic cities: Integrating nature into urban design and planning*. Island Press.

Chan, F. K. S., Li, L., Cheshmehzangi, A., Thadani, D. R., & Ives, C. D. (2022). The transformation of the green infrastructure intervention under the case of Sponge City Program: Positions, challenges, and prospects in China. *Green Infrastructure in Chinese Cities*, 97–122.

Cheshmehzangi, A., Alidoust, S., Chen, W. Y., & Fuller, R. (Eds.). (2026a). *Nature in cities, nurturing cities*. Springer.

Cheshmehzangi, A., Alidoust, S., Chen, W. Y., & Fuller, R. (Eds.). (2026b). *Nature-based urbanism*. Springer.

Cook, I. R., & Swyngedouw, E. (2012). Cities, social cohesion and the environment: Towards a future research agenda. *Urban Studies, 49*(9), 1959–1979.

Farr, D. (2011). *Sustainable urbanism: Urban design with nature*. Wiley.

Gentin, S., Pitkänen, K., Chondromatidou, A. M., Præstholm, S., Dolling, A., & Palsdottir, A. M. (2019). Nature-based integration of immigrants in Europe: A review. *Urban Forestry & Urban Greening, 43*, 126379.

Grace, M., Scott, A. J., Sadler, J. P., Proverbs, D. G., & Grayson, N. (2023). Exploring the smart-natural city interface; re-imagining and re-integrating urban planning and governance. *Emerald Open Research, 1*(5).

Johnson, E. A., & Klemens, M. W. (2005). The impacts of sprawl on biodiversity. In *Nature in fragments: The legacy of sprawl* (pp. 18–54). Columbia University Press.

Li, L., Chan, F., & Cheshmehzangi, A. (2023). Nature-based solutions and sponge city for urban water management. In *Adapting the built environment for climate change* (pp. 371–402). Woodhead Publishing.

Li, L., Cheshmehzangi, A., Chan, F. K. S., & Ives, C. D. (2021). Mapping the research landscape of nature-based solutions in urbanism. *Sustainability, 13*(7), 3876.

Li, L., Ives, C. D., Carter, J., Collins, A. M., Adekola, O., Green, D., Cheshmehzangi, A., & Chan, F. K. S. (2026). Examining Sponge City coverage and perceptions in global news media. *Cities, 169*, 106588.

Paniagua-Villada, C., Garizábal-Carmona, J. A., Martínez-Arias, V. M., & Mancera-Rodríguez, N. J. (2024). Built vs. Green cover: An unequal struggle for urban space in Medellín (Colombia). *Urban Ecosystems, 27*(4), 1055–1065.

Postma, D. W. (2006). *Why care for nature? In search of an ethical framework for environmental responsibility and education*. Springer Netherlands.

Ren, Q., He, C., Huang, Q., Zhang, D., Shi, P., & Lu, W. (2023). Impacts of global urban expansion on natural habitats undermine the 2050 vision for biodiversity. *Resources, Conservation and Recycling, 190*, 106834.

Simeonova, V., & Van Der Valk, A. (2016). Environmental policy integration: Towards a communicative approach in integrating nature conservation and urban planning in Bulgaria. *Land Use Policy, 57*, 80–93.

Simkin, R. D., Seto, K. C., McDonald, R. I., & Jetz, W. (2022). Biodiversity impacts and conservation implications of urban land expansion projected to 2050. *Proceedings of the National Academy of Sciences, 119*(12), e2117297119.

Sun, J., Cheshmehzangi, A., & Wang, S. (2022). Comprehensive evaluation of green infrastructure restorative practices for high-quality transitional "sponge node" renewal programs in China. In *Green infrastructure in Chinese cities* (pp. 265–298). Singapore.

Williams, D. E. (2007). *Sustainable design: Ecology, architecture, and planning*. Wiley.

Wong, G. (2024). Lion City zoopolis: Urban crittizenship in biophilic Singapore. *Anthrozoös, 37*(2), 213–229.

Design Methods and Material Practices

Nature-Built: Greening Tomorrow's Subtropical Cities with Native Plants

Silvia Micheli, Susanne Schmidt, and Lu Jin

Abstract Nature is becoming integral to urban development, with cities enhancing liveability and climate control by incorporating plants. Yet, urban projects often have lengthy and complex timelines, and green elements are typically addressed last, making them vulnerable to budget cuts and design compromises. By integrating plants from the outset as a core design element, their inclusion boosts effectiveness. In subtropical Australia, living walls are emerging as key features of contemporary architecture and urban greening strategies. These installations often rely on a limited range of tropical species, inspired by global models like Singapore and Ho Chi Minh City.

While visually appealing and resilient, these tropical plants may not be ideal for regions like subtropical Queensland that have its own rich native flora. This chapter explores plant integration in building design, emphasising native species as more ecologically appropriate and supportive of local biodiversity and identity. We investigate living walls in Brisbane through case studies, highlighting their role as micro-landscapes that enhance urban resilience. We then explore how selecting suitable native plants adapted to subtropical climates can boost ecological health to create place-specific aesthetics. This chapter also addresses risks, such as vulnerability to extreme weather, pest issues, and structural concerns such as water infiltration and maintenance. Still, the ecological advantages of native plants outweigh these challenges, offering greater resilience for subtropical cities among other benefits. Drawing on field investigations and comparative studies, we propose a preliminary framework for advancing living wall design as a tool for tomorrow's sustainable urban futures.

Keywords Living wall design · Place-specific design · Native plants · Green infrastructure · City resilience · Biodiversity · Productive city

S. Micheli (✉) · L. Jin
School of Architecture, Design and Planning, University of Queensland, Brisbane, Australia
e-mail: s.micheli@uq.edu.au

S. Schmidt
School of Agriculture and Food Sustainability, University of Queensland, Brisbane, Australia

A. Cheshmehzangi et al. (eds.), *Designing Nature into Cities*, Urban Sustainability, https://doi.org/10.1007/978-981-95-9382-8_2

1 Introduction: Scene Setting for Living Architecture in the Asia-Pacific Region

The capital city of Queensland, Brisbane, is recognised as a relevant subtropical case for greening practices in the subtropical urban setting. In this chapter, Brisbane is analysed in the broader context of the Asia-Pacific Region as a testing ground for the integration of native plants in the buildings' architectural design process. After a survey of how plants are increasingly considered in the design of urban buildings in countries like Singapore and Vietnam, research questions related to the use of living elements underpin a new methodology here advanced for effectively integrating plants in the architectural design process. This methodology sits in the context of two case studies relevant to the Brisbane contemporary design scene. Finally, design features of selected native plants were identified to understand how they appropriately match their particular subtropical climatic conditions.

In the Asia-Pacific region, green architectural elements, including living walls and vegetated grounds, facades, columns, trusses, and roofs have become an integral part of urban greening strategies, advancing contemporary cities' sustainable development agendas and sustainability transitions (Bush, 2020). In the late twentieth century, many urban centres in the region adopted Western constructive models, embracing concrete and curtain-wall systems, with little regard for local environmental conditions or building traditions. This trend, while indicative of a desire of renewal, resulted in buildings that underperformed climatically and disconnected from local practices. In recent years, the Vietnamese government has supported private ownership of land, setting the administrative conditions for unique design opportunities (O'Connell, 2018). In countertendency to the thesis formulated in 2001 by Tay Kheng Soon, according to which 'Western hegemony and the dominance of its architectural and artistic media have significantly contributed to the non-Western nations' architects dampening of their own ability to respond to their own climatic and cultural imperatives as sources for innovation' (Tay, 2001), a new generation of local architects has undertaken to re-engage with and advance the local building tradition, considering plants an essential element of design (Micheli, 2020).

Thus, a paradigmatic shift is occurring. As cities such as Singapore, Ho Chi Minh City and Brisbane experience rapid economic growth, they are reconfiguring their urban planning strategies and architectural design processes to embrace local climatic advantages (Micheli & Brugman, 2023). A key component of this transformation is the deployment of nature-integrated surfaces—or 'living green facades'—as vital architectural tools in redefining the tropical urban function and aesthetic. These vertical gardens, often arranged with climbing plants and supported by irrigation or hydroponic systems, not only serve as filters reducing urban dust, noise, and glare, but also act as symbolic markers of sustainability and affluence. They represent a new 'green paradigm' (Knill, 1991) where the boundary between the building and nature is deliberately blurred. High-rises, once energy-intensive monoliths, are now envisioned as 'green towers' that breathe, shade, and cool.

As such, the green wall is no longer an aesthetic addition; it is a strategy in densifying cities where ground-level green space is scarce. It contributes to an alternative urban scenography where lush verticality compensates for horizontal spatial constraints. The living green façades become a loose divider between architectural and urban spaces, thus acting as both a climatic tool and a rhetorical device—'living curtains' that mask raw structures while transforming them into ecological statements (Micheli, 2016).

As cities strive to become 'cities in a garden', the role of living facades becomes increasingly central—not just in mitigating environmental impacts but in shaping a renewed urban identity that is regionally grounded. Applied to the subtropical region of Queensland in Australia though, this typology requires further studies in the attempt of deploying native plants for more performant results.

A paradigmatic example of this approach is WOHA's Oasia Downtown Hotel in Singapore, where plants are fully integrated into the building envelope, embodying a seamless transition between interior and exterior. The favourable tropical climate allows for porous, open designs that replace the need for sealed envelopes, embracing airflow, light, and outdoor connectivity. The hotel is wrapped by a double skin, whose external part is a red metallic mesh. This material allows the philodendron and companion tropical plants to hook onto the support and run onto it (see Fig. 1a,b). While the architect thought of a single species, the final result is a combination of plants to give the variety necessary to mitigate the risks of issues related to one species, like the presence of parasites.

Fig. 1 **(a, b)** Oasia Downtown Hotel, Singapore. Details of the façade system, with growing plants. Design by WOHA (Photos taken by the authors)

Similarly, but at a different scale, the Breathing House in Ho Chi Minh City by Vo Trong Nghia Architects (VTN) exemplifies a refined approach to tropical urban living, using natural elements functionally rather than decoratively (Belogolovsky, 2021). Immersed in a narrow site in an alleyway, the residence is wrapped by a 'green veil', thought as an 'environmental diffuser, filtering harsh tropical sunlight, improving the view while ensuring security and privacy' (Vo Trong Nghia, 2019). This experimental architecture maintains the essence of Asian architecture (Vo Trong Nghia, 2015) while becoming a transferable example of urban sustainability. Rare in architectural drawings, the selected species for the project are represented in the section drawing of the building, demonstrating VTN's understanding of plants as a living material. *Vernonia elliptica*, a curtain creeper, and *Combretum indicum*, a woody vine, are combined as plants that grow in different directions: the former gently cascading down and the latter climbing up the building. These two directions are favoured by the support provided by the building. The two species populate the double mesh metallic facade, which allows for natural cross-ventilation and privacy for the family without hindering the filtering of natural light. On this support, plants thrive also because they absorb the humidity in the air. In buildings like that, the facades operate as transitional elements (Tay, 2001), and the emphasis is put on the indoor experience of the space and natural elements. Although individual dwellings have a minor visibility in city-making processes, their influence can be significant culturally, as their example can catch the attention of the authorities and inform new approaches to urban planning (O'Connell, 2018).

The Jakob Factory, designed by G8A Architecture & Urban Planning in collaboration with Rollimarchini Architekten, is a notable example of innovation in the design of living surfaces. In this project, a wide array of tropical plants is integrated into the building's facade through long, suspended planting trays, which run horizontally for the whole light of the facades showcasing the strength of the stainless-steel cables produced on-site (see Fig. 2). While the system successfully contributes to cooling the interior, the ongoing gardening required to maintain and contain the plant growth adds to the building's overall maintenance costs.

These international case studies show the growing interest and expertise in using plants as living materials integrated in the architectural design process. However, these and other examples are mostly located at the tropics, with a specific selection of plants proper to that latitude.

2 Research Questions for Future Directions for the Design of Living Architectural Elements

Despite the growing popularity of living walls in contemporary architecture, these systems often rely on a narrow selection of plant species that conform to a standardised global green aesthetic based on the performant tropical examples. This aesthetic is heavily influenced by Singapore and Vietnam's internationally

Fig. 2 The Jakob Factory, Ho Chi Minh City, Vietnam. Design by G8A Architecture & Urban Planning (Photos taken by the authors)

recognised greening programmes which promote the use of tropical species known for their resilience, ease of maintenance, and visual appeal. However, the transplantation of this model into ecologically distinct regions—such as subtropical Queensland—raises questions about its relevance and long-term sustainability. There are also considerations regarding building and maintenance-related costs, which differ from country to country.

This chapter investigates the potential for integrating native plant species into design elements, arguing that locally adapted flora offers ecological, cultural, and functional advantages. By aligning plant selection with regional biodiversity goals and environmental conditions, living walls can move beyond superficial greening to become meaningful contributors to specific urban ecosystems. Underpinning this inquiry of evidence-based research is the recognition that the integration of vegetation into architectural systems is a critical urban issue for the sustainability of our cities. Central to the research are the following questions:

1. How can architecture respond to and support the dynamic behaviour of plants?
2. What are the specific requirements, risks, and challenges associated with native subtropical species?

Plants are considered in this study as building materials—responsive, dynamic agents whose design implications extend beyond aesthetics or technical performance. This perspective addresses a gap in architectural discourse, where vegetation is often approached from a purely decorative or technological standpoint and in relation to non-site-specific ecologies. By rethinking plant integration as a responsive, ecologically grounded design process for subtropical climates, this research calls for a departure from imported models and advocates for site-specific frameworks that reflect local biodiversity and climatic realities. Understanding subtropical plant behaviour and developing architectural frameworks that accommodate this complexity is essential—not only for the success of green infrastructure, but for the broader project of aligning urban development with ecological resilience.

3 Methodology for Integrating Plants in the Architectural Design Process

The originality of this study lies in its interdisciplinary exploratory approach, where architecture and horticulture are interpolated to investigate the direct relationship between the living component (plants) and the inert materials in the design and construction process. The aim is to study the interactions resulting from the integration of plants and the building.

To explore the integration of subtropical plants into the design of architectural elements—with attention to native species in Queensland—this study develops a methodological model grounded in an interdisciplinary and design-oriented approach applied to case studies. The approach unfolds across several stages.

1. The process begins with the development of a plant species inventory for each site. Species are identified through the collection of physical specimens and detailed photographic documentation. Where available, official species lists are sourced by consulting the project's architects or consultants. The selection is conducted factoring in design needs, risks and challenges associated with green design.
2. Once the species list is established, plant characteristics are documented. Categories of typical plant traits according to botanical and ecological standards are identified in a matrix, ensuring consistency across cases and allowing for comparison. This table is called Matrix of Plant Characteristics (see Table 1). The Philodendron is taken as an example for its common application in architectural design.
3. Building on this, a parallel analysis addresses how plant characteristics interact with and are supported by architectural design. This involves mapping out some the plants' design-specific qualities—such as spatial arrangement, material interface, and environmental responsiveness—highlighting the ways in which the built structure accommodates or enhances vegetative growth. This second

Table 1 Matrix of plant characteristics (Table by the authors)

Property	Details
Botanical name	*Philodendron* species
Plant type	Tropical perennial (often grown as a houseplant)
Family	Araceae
Light requirements	Bright, indirect light (some tolerate low light)
Watering needs	Moderate
Soil type	Well-draining potting mix (rich in organic matter)
Humidity	Prefers high humidity
Temperature range	18–27 °C; avoid cold drafts
Growth habit	Climbing or trailing (vining), some upright varieties
Leaf appearance	Heart-shaped, lobed, or deeply cut; glossy green leaves
Toxicity	Toxic to pets and humans if ingested (contains calcium oxalate crystals)
Propagation	Typically by stem cuttings
Common varieties	*Philodendron hederaceum, P. selloum, P. gloriosum, P. micans*, etc.
Maintenance level	Low to moderate; relatively easy to care for
Common issues	Root rot, leaf yellowing (from overwatering), pest infestations (mealybugs)

Table 2 Matrix of plant characteristics for the design process (Table by the authors)

Property	Details
Growth habit	Herbaceous, climbing, hanging
Growth reproduction	Invasive/non-invasive (depending on species)
Support needs	Requires a structure like moss poles, trellises, walls, or stakes to climb
Aerial roots	Present; used to anchor onto rough surfaces like bark or textured walls
Light and humidity requirements	Prefers bright, indirect light
Seasonal patterns	Evergreen
Availability of stock	Commonly used in urban settings. Easily available for purchase at plant nurseries
Substrate requirements	Consistent moisture
Impact on surfaces	Potentially damaging; roots attach to climbing surface, carry moisture and may act as a conduit for insects
Growth rate and lifespan	Moderate to fast under ideal conditions; lifespans range from short-lived to decades
Plants interaction and habitat	Coexists well with other species, can create shaded microclimates
Types of ecosystem services	Biodiversity, air filtration, not edible for humans
Maintenance possibilities and requirements	Needs to be shaped and pruned regularly

matrix is called the Matrix of Plant Characteristics for the Design Process (see Table 2). The Philodendron is again taken as an example.
4. To deepen the spatial understanding of each project, photographic documentation is employed. Particular attention is given to the moments of physical contact between plants and the architectural surface and structure. These images serve as critical visual data for interpreting the lived experience of the design.
5. Beyond the visual and material aspects, the model calls for an investigation into the design intent behind each vegetated wall. This includes uncovering the criteria for plant selection, the intended benefits—whether ecological, aesthetic, or experiential—and the broader design goals the vegetation is intended to support.
6. The model concludes with an evaluation of the risks and challenges inherent to integrating vegetation into architectural systems. This includes considerations of technical limitations, environmental stressors, and the long-term viability of plant-based components within built environments.

By synthesising horticultural knowledge with design analysis, this methodological model offers a preliminary framework for understanding how vegetation is strategically embedded with the building. Recognising that green infrastructure evolves over time, the methodology considers a temporal dimension. This involves examining plants' lifespan, propagation cycles, and maintenance routines(Sample, 2016; Glazebrook et al., 2022). Both successes and unexpected outcomes are analysed to assess the resilience and adaptability of the integrated design approach.

4 Comparative Case Studies

As 'explorations of built examples are rare' (Weijie Zhong et al., 2023), this research has selected two case studies in the local urban environment of the subtropical city of Brisbane for which in-person site visits have been conducted. The entry veranda of the Terrarium House (2017) by architect John Ellway and the hanging FytoColumn installation in the Queen's Wharf redevelopment (2024) by Cottee Parker/Fytogreen (ecological design consultant) have been chosen for their innovation in integrating plants as crucial architectural design elements.

Terrarium House, a demonstration project that showcases Ellway's innovation in developing a place-based living design, adds to the city's liveability and subtropical landscape by the means of its facade (see Fig. 3a,b). According to the jury that awarded the project with the Brisbane 2018 'Houses Award: House Alteration and Addition', 'Ellway has deftly exploited the site's natural fall by inserting living spaces underneath the raised original cottage. This move has deconstructed the classic veranda arrangement to create a wonderful, breezy and vine-covered entry and external vertical circulation. The design of the Terrarium House is attuned to its subtropical climate, promoting cross-ventilation with a sunken living space protected from the harsh Queensland sun'.

The critical first step in evaluating the integration of vegetation within this residential project was the making of a plant species list. The current plants include a

Fig. 3 (**a, b**) Terrarium House, inner-city Brisbane, Queensland. Design by John Ellway Architect (Photos taken by the authors)

mixture of native plant species (including *Cissus antarctica* 'Kangaroo Vine' and *Atractocarpus fitzalanii* 'Native Gardenia') and non-native plants commonly used as indoor plants (including *Monstera deliciosa* 'Swiss Cheese Plant', *Syngonium podophyllum* 'Arrowhead Vine', *Epipremnum aureum* 'Pothos', *Ficus pumila* 'Creeping fig', and several fern species).

Although *Passiflora edulis* 'Panama Gold' passionfruit had been planted in the outset as the predominant climbing plant with a fast growth rate for the facade, it was gradually replaced by Kangaroo Vine. Reasons for this include that passionfruit plants tend to lose leaves and can at times appear unkempt; the ripening passionfruit attract possums who would leave behind scat; the sweet sap on the stems attracted ants. Kangaroo Vine on the other hand had characteristics that are more appropriate for this function and was selected for further analysis (see Table 3).

The design intent of the project, as the name of the residence implies, is that the structural integration of the plants is central to the architectural strategy. Vegetation is supported by two built components: an off-the-shelf metal mesh and a retaining wall. The mesh, constructed from 100×100 mm, 5 mm-thick wire, acts not only as a climbing structure for the vines but also as balustrading for the staircase and a bridging walkway, thus merging botanical and architectural elements into a unified system. The interaction between plant morphology and structural form is deliberate; Kangaroo Vine is particularly well-suited to this type of support system, enabling vertical growth while offering flexibility in terms of coverage and density. According to the architect, while the plant selection reflects a relatively modest palette, this simplicity was intentional. For instance, although flowering species such as *Jasminum officinale* 'Common Jasmine' were considered, they were ultimately excluded due to concerns over strong fragrances and clients' personal preferences. The result is a simplified composition with one dominant vine, chosen for ease of maintenance and visual consistency.

Table 3 Matrix of plant characteristics for the design process of the Kangaroo Vine (*Cissus antarctica*) in the Terrarium House (Table by the authors)

Property	Details
Growth habit	Vigorous evergreen climber, large glossy green leaves
Growth reproduction	Native to eastern Australia; seed germination is slow and unpredictable, easily propagated from cuttings
Support needs	Can grow 3–6 m or more if supported by vertical structures, using tendrils to grasp surfaces
Aerial roots	Not present
Light and humidity requirements	Prefers bright, indirect light
Seasonal patterns	Evergreen
Availability of planting stock	Relatively common in urban settings; easy to source
Substrate requirements	Tolerant of a variety of soil types
Impact on surfaces	Minimal
Growth rate and lifespan	Moderately slow to establish but can grow quickly under ideal conditions once established. Lifespan can be decades
Plants interaction and habitat	Not applicable, as it is the only plant species on the vertical plane
Types of ecosystem services	Local biodiversity, the acidic berries can be eaten raw or cooked; air filtration
Maintenance possibilities and requirements	Requires regular pruning during warmer months due to vigorous growth, less maintenance required during cooler months

Understanding the design intent further clarifies the rationale behind the plant selection in relation to spatial organisation. One primary goal was visual privacy. Kangaroo Vine serves to obscure views into the interior of the house, particularly at the upper levels, where foliage is densest. In contrast, lower sections of the wall, where the vine stems become woody and sparse over time, maintain a degree of openness. This allows residents to retain visual connection with the street while benefiting from privacy within the home—the design thus achieves a balance between seclusion and passive surveillance. Security considerations also shaped material choices. The off-the-shelf mesh was preferred over vertical cables or other mesh systems due to its availability, cost, as well as robustness, allowing large doors and windows to remain open during the day without compromising safety.

Beyond privacy and security, the green wall serves several environmental functions. The leaf surface area captures dust particles from the street, reducing particulate matter entering the house. This filtration effect is tangible; the plants require regular washing, providing a visible record of the pollutants they intercept. Additionally, the green wall moderates the local microclimate. During storms, the dense vegetation deflects rain, sheltering the stairwell and garden. In summer, watering the foliage produces an evaporative cooling effect, helping to reduce temperatures in adjacent living spaces.

At present, some plants face challenges with pests, particularly mealybugs. These insects secrete a sticky substance that attracts ants and facilitates the growth of sooty mould, impacting both the health and appearance of the vegetation. Such issues highlight the need for careful species selection based on site-specific conditions and ecological resilience.

The Queen's Wharf project presents one of several vertical garden systems developed by Fytogreen that combine structural engineering with ecological design to support plant life on building surfaces. The FytoColumn system uses suspended cables to carry modular donut-profiled engineered foam substrates that serve as lightweight, porous growing elements (see Fig. 4a,b). Irrigation tubes drip-feed water and nutrients to the top of each foam segment, allowing the liquid to percolate downward for plant roots to access. A pump at the base of each column collects and recycles excess liquid to the top, creating a closed-loop watering system.

In contrast to the simplicity of the Terrarium House, this system incorporates 82 plant species (and 1890 individual plants), suspended within a high-rise atrium with highly specific micro-climatic conditions, with critical factors including wind and aspect.

While the design intent is clear, having a scientific understanding of how a plant grows in its native environment, its adaptive strategies, how it reproduces, and how it interacts with other species over time is essential to this project's ongoing success. Plants used include epiphytes and lithophytes (plants that naturally grow on other plants or rocks) as well as vines, grasses, ferns, and even fig trees that are well adapted due an aerial life cycle. The design aims to create a biodiverse system not only in species count but in ecological function and occupation; a surface with only *Lomandra longifolia* (a common grass-like native plant), or even multiple species

Fig. 4 (**a**, **b**) Queen's Wharf in Brisbane's CBD. Design by Fytogreen/Cottee Parker (Photos taken by the authors)

with identical growing habits, would still be considered a monoculture and not fulfil the biodiversity and resilience criteria. Furthermore, since each plant species will change in size and number due to its natural growth, maintenance interventions, and ecological succession, how these aspects will impact the design requires specialised considerations.

Despite the need for highly technical ecological knowledge to be integrated into the architectural and structural design, Fytogreen is usually brought into projects after design and budget decisions have already been made between architects and clients, thus limiting their influence. Established building construction processes means the project timeline and budget allocated for plants may limit the species selection—some native plants may have been ideal in theory but unavailable in nurseries, too slow-growing to be viable, or too expensive to acquire. Labour costs in Australia are also high relative to many of its Asia-Pacific counterparts, leading clients to prefer low-maintenance solutions to minimise ongoing costs.

Furthermore, the aesthetic requirements of a project are typically established between the client and architect, so there are limits to Fytogreen's influence on the broader aesthetic. For example, when the architect used a render of a leafy, tropical green installation design, clients expected plants selected to fit within this aesthetic profile; colourful flowering plants would not meet such expectations. However, given that many green walls are not viewed from up-close, incorporating small flowering plants that are not perceivable from the street level would provide food and habitat for birds and insects. Nevertheless, this aspect wasn't necessarily a priority on the client's agenda.

5 Design Features of Selected Native Plants for Living Elements

In this chapter, a range of native plant species were selected for discussion to better understand how they appropriately match their particular subtropical climatic conditions. Drawing on the comparative analysis of the case studies for selecting plants allows a deepened understanding of their ecological fit. This supports the approach advocated here for the integration of plants through design, one that emphasises plants native to local subtropical environments as agents of enrichment of ecological health as well as vehicles to create a place-specific aesthetic in a distinct environmental setting.

The characteristics of native plants underline their usefulness in local contexts, being largely common resilient houseplants, easy to propagate, and maintain. The chapter has discussed the fit of the selected plants to architectural structures as living materials, noting the service they offer to the built environment, including water retention and evaporative cooling to help mitigate the urban heat island effect; shading buildings to reduce energy consumption and counteracting the effects of heatwaves. Air purification by trapping particulate matter is a further benefit, thus improving air quality and alleviating respiratory conditions. Native living walls can support urban biodiversity (Berthon et al., 2021), from microbial organisms to invertebrates and mammals.

As a design consideration, risks and challenges associated with green design need to be considered in the plant selection. These include the potential vulnerability of plants to extreme weather events (such as cyclones), seasonal-related behaviours of plants; extremes in temperature, droughts, and fires; the possible negative impacts on human living conditions, such as the attraction of organisms that carry disease (for example, mosquitoes) or reach plague proportions (e.g., brushtail possums). Additionally, structural risks related to water infiltration, excessive mould growth, material decay, and long-term maintenance requirements are considered. Time is a factor too in architectural design through the consideration of patterns of plants growth, which poses another layer of complexity in the deployment of the living material reacting to the environment. However, these risks are to be considered against the ecological gains, including (1) increased biodiversity for butterfly habitat; (2) increased biodiversity for bees; (3) increased production of food for humans (e.g., wombat berries, scrambling lily); and (4) water saving provided by using native plants, which provide a more performant response for improved place-based resilience.

6 Conclusions

By integrating insights from both fields of investigation—architecture and horticulture—and through evidence-based research, this chapter has provided a preliminary framework for framing the use of living elements in architectural design as tools for urban sustainability in subtropical cities. In doing so, it advances site-specific design paradigms and strategies to enhance native-plants integrated design in the urban environment.

This chapter has discussed the role of native plants in living element systems, considered as micro-landscapes within the urban fabric of cities. In looking at the species selection process and design integration, this approach addresses gaps in the current research about green architecture and its design integration, offering at the same time a local solution for urban greening, and a transferrable method for urban resilience.

This approach should inform and be supported by policy guidelines and the development of industry standards that mandate the integration of site-specific ecological knowledge early in the architectural design process, ensuring long-term sustainability and place-based responsiveness in the urban development of tropical and subtropical cities.

Acknowledgements We would like to acknowledge the kind support and assistance of architect Nguyen Duy Tan from G8A Architects for the site visit at the Jakobs Factory in Ho Chi Minh City, 2024; John Ellway for the site visit at the Terrarium House in Brisbane, 2025; and Erik van Zuilekom from Fytogreen for the site visit to the hanging FytoColumns at Queen's Wharf in Brisbane, 2025.

References

Belogolovsky, V. (2021). *Green architecture: Vo Trong Nghia & the work of VTN Architects*. Prestel.

Berthon, K., Thomas, F., & Bekessy, S. (2021). The role of 'nativeness' in urban greening to support animal biodiversity. *Landscape and Urban Planning, 205*, 103959. https://doi.org/10.1016/j.landurbplan.2020.103959

Bush, J. (2020). The role of local government greening policies in the transition towards nature-based cities. *Environmental Innovation and Societal Transitions, 35*, 35–44. https://doi.org/10.1016/j.eist.2019.04.002

Glazebrook, J., Leiper, G., & Cox, D. (2022). *Mangroves to mountains: A field guide to the native plants of South-east Queensland* (3rd ed.).

Knill, G. (1991). Towards the green paradigm. *South African Geographical Journal, 73*(2), 52–59. https://doi.org/10.1080/03736245.1991.9713549

Micheli, S. (2016). Tropical facade: A manifesto for the Asia Pacific region. In *The Singapore Architect* (pp. 61–69).

Micheli, S. (2020). The ecology of (sub)tropical residential architecture in urban Vietnam and Australia. Urban Tropicality: 7th International Network of Tropical Architecture Conference, Brisbane, Australia, 5–8 December 2019. Brisbane, Australia: The University of Queensland.

Micheli, S., Brugman, J. (2023) 'The Changi-Marina Bay Corridor: green strategies for Singapore's soft power', Architectural Research Quarterly, 27, (pp. 61–71). https://doi.org/10.1017/S1359135522000598.

O'Connell, R. (2018, March 17). How Ho Chi Minh City architects are reshaping the city, in the spirit of Zaha Hadid. *Post Magazine*. https://www.scmp.com/magazines/post-magazine/long-reads/article/2137270/how-ho-chi-minh-city-architects-are-reshaping

Sample, H. (2016). *Maintenance architecture* (1st ed.). MIT Press. https://doi.org/10.7551/mitpress/9316.001.0001

Tay, K. S. (2001). Rethinking the city in the tropics. In E. Viray (Ed.), *The Asian everyday* (p. 299). TOTO.

Vo Trong Nghia. (2015). Save our Earth. In E. Viray (Ed.), *The Asian everyday* (p. 116). TOTO.

Vo Trong Nghia. (2019). Breathing house. *GA Houses, 162*, (p. 113).

Weijie, Z., Schroeder, T., & Bekkering, J. (2023). Designing with nature: Advancing three-dimensional green spaces in architecture through frameworks for biophilic design and sustainability. *Frontiers of Architectural Research, 12*(4), 732–753. https://doi.org/10.1016/j.foar.2023.02.005

Material Paradox in Nature-Driven Placemaking: Contested Sustainability at Barongan Market, Indonesia

Muhammad Nelza Mulki Iqbal

Abstract This chapter examines the contested narratives of sustainability at Barongan Market, a temporal marketplace in peri-urban East Java, Indonesia. While publicly promoted as eco-conscious placemaking initiatives, its everyday material practices—such as the re-emergence of plastics and reliance on non-local 'natural' packaging—reveal a disconnection between symbolic sustainability and ecological realities. Drawing on Materiality and Assemblage Theory, this chapter conceptualises materials like bamboo, banana leaves, paper, and plastic not as passive objects but as active agents shaping economic behaviour, ecological discourse, and community adaptation. Based on ethnographic fieldwork, the findings suggest that sustainability at Barongan Market is often commodified, shaped more by aesthetic branding and economic pressure than by systemic ecological transformation. The chapter contributes to debates on temporal placemaking in the Global South context by showing how sustainability is performed, negotiated, and often contradicted through material paradoxes inherent in its assemblage. It calls for place-sensitive planning and design that moves beyond green branding towards grounded accountability and support for local material ecosystems.

Keywords Temporal markets · Sustainability contradictions · Materiality · Assemblage theory · Placemaking · Southern urbanism

M. N. M. Iqbal (✉)
School of Architecture and Landscape University of Sheffield, Sheffield, UK

Institut Teknologi Nasional Malang, Malang, Indonesia
e-mail: mnmiqbal1@sheffield.ac.uk

A. Cheshmehzangi et al. (eds.), *Designing Nature into Cities*, Urban Sustainability, https://doi.org/10.1007/978-981-95-9382-8_3

1 Introduction

In the Global South, the concept of nature is not fixed, but is shaped by unique historical, social, and cultural contexts (Droz et al., 2022, 2023). This perspective contrasts with dominant Western binaries—such as nature versus culture or human versus non-human—which often portray humans as separate from or threatening to nature (Anderson et al., 1996; Plumwood, 2019). Instead, the Global South provides a critical lens to examine the entanglements of knowledge, power, and urban transformation (Mady et al., 2025; Shin, 2021). Understanding nature through this lens requires attention to temporalities, plural worldviews, and the risks posed by universalised conservation narratives, which can lead to adverse outcomes like dispossession (Dovey et al., 2019). Nature must therefore be understood as relational and evolving, embedded in the political, social, and material fabric of everyday life.

Peri-urban areas in the Global South are often sites of hybridity and experimentation (Rajendran et al., 2024). In Indonesia, these conditions are reflected in *desakota* regions—fluid blends of village and urban forms (McGee, 1989; Sui & Zeng, 2001). Within these settings, temporal placemaking practices—such as short-term markets—emerge as strategies supporting social, ecological, and economic goals (Glazener et al., 2022). While often celebrated for promoting community well-being and environmental awareness, their material practices remain underexplored.

This chapter addresses that gap by critically examining the material paradox embedded in Barongan Market, a temporal market renowned for its eco-conscious image. While publicly identified with its natural aesthetic, its everyday practices tell a more complex story: the reappearance of plastic, reliance on imported organic packaging, and inadequate waste infrastructure challenge its sustainability claims. To unpack these tensions, the study draws on Materiality and Assemblage Theory (Deleuze & Guattari, 1987; Latour, 2005), viewing materials not as static site but are dynamic agents that co-constitute social, economic, and ecological processes. This lens prompts an inquiry not into success or failure of the materiality, but into the relational processes that constitute it. Therefore, this chapter asks:

1. How do the materials in the Barongan Market support or contradict its sustainability branding?
2. How do everyday material practices shape perceptions and experiences of the market?

By centring these questions, the chapter challenges aestheticized and symbolic framings of eco-placemaking practice. It argues that understanding sustainability in the Global South setting like what happen in Indonesia requires a grounded, material-sensitive approach—one that is accountable to the complex assemblage of local dynamics and community capacities. The next section situates this study within conceptual debates on nature, sustainability, and placemaking.

2 Nature, Sustainability, and Placemaking in the Global South

The 'Global South' operates as a critical lens that challenges Eurocentric urbanism and advances alternative epistemologies that recognise hybrid urban forms, formal-informal coexistence, and compressed temporalities (Wood, 2020; Mady et al., 2025). These hybrid conditions create complex assemblages that challenge simple definitions of sustainability. This "Southern palimpsests" reflects overlapping dynamics, prompting a rethinking of agency, participation, and resource flows. While some scholars note the empirical distinctiveness of Southern cities, marked by irregular governance, others emphasise their experimental and innovative characters (Shin, 2021; Schindler, 2017; Wood, 2020; Caprotti & Cowley, 2017). In these dynamic contexts, the relationship between nature, sustainability, and placemaking becomes a crucial entanglement to explore.

Nature itself is a multifaceted concept (Burns & Witoszek, 2012; Ducarme & Couvet, 2020). Western discourse often renders it separate from human activity, marginalising relational conceptions grounded in integration and interdependence. In many Indigenous and Asian cosmologies, nature is embedded within spiritual, ancestral, and communal domains, revealing how human–environment interconnectedness is culturally produced (Droz et al., 2022). This calls for pluralism in ecological thought to avoid universalising blind spots (Hakkarainen et al., 2020).

Sustainability discourse similarly traverses multiple registers. Initially defined by the 1987 Brundtland Report (Burns & Witoszek, 2012; Keeble, 1988), the concept remains criticised for marginalising its social dimension (Kuhlman & Farrington, 2010; Salama et al., 2024). Griep et al. (2023) argue that urban sustainability often sidelines the "people" component, favouring technocratic narratives. This critique is extended by Beasy & Corbett (2021), who argue that mainstream sustainability is a discourse shaped by urban, middle-class perspectives. Addressing these imbalances requires acknowledging the spatial and social inequalities that affect how communities sustain their environments (Salama et al., 2024; Iqbal & Shafique, 2025). Building on this socio-spatial critique, I adopt people–place narratives as an analytical lens: the lived attachments, routines, and neighbourhood relations through which social sustainability becomes legible.

Placemaking itself has gained increasing scholarly attention since for its role in fostering social cohesion, preserving heritage, and imagining sustainable futures (Yudistira, 2017; Yan et al., 2024). Previous research shows how temporary markets in Indonesia's desakota regions, like Barongan Market, can sustain local heritage and resist capital-intensive development through community-led, temporal urbanism (Iqbal & Shafique, 2025). Yet, this intersection of nature, sustainability, and placemaking invites critiques. Droz et al. (2024) warn against eco-orientalism—idealised portrayals of Asian environmentalism—while others highlight the need for more relational approaches that address embedded hierarchies (Nielsen, 2004; Plumwood, 2019).

Southern urbanism offers a vital framework for understanding sustainability as a situated, relational, and contested process (Mady et al., 2025). By emphasising temporal negotiation and plural epistemologies, it positions the Global South as a generative space for reimagining urban futures through its diverse placemaking practices.

3 Materiality and Assemblage

Recent scholarship has reinterpreted architectural and urban phenomena through assemblage thinking, a framework from Deleuze and Guattari (1987). Assemblages are dynamic compositions of material, social, spatial, and expressive elements, connected through emergent relations (Nail, 2017). This perspective challenges static binaries like formal/informal or top-down/bottom-up, offering a more fluid understanding of socio-spatial complexity (Shafique, 2022). Applied to Barongan Market, it reveals not just a temporal, eco-themed site, but a space shaped by relational and contested material networks.

Assemblage theory allows us to understand urban and architectural phenomena as emergent from networks of diverse elements—not limited to human actors, but extending to technical, material, environmental, and conceptual forces. Deleuze and Guattari's foundational text, "A Thousand Plateaus", describes assemblages as both mechanical (material) and enunciative (expressive) (Deleuze & Guattari, 1987). Deleuze speaks of thinking in rhizomes, referring to roots that grow interconnected in various directions. However, assemblage theory is often misread as a mere act of "assembling", thus stripping it of its ontological richness (Buchanan, 2017; Nail, 2017). In response, McFarlane and Anderson (2011) propose that assemblage should be treated not only as a descriptor or analytic tool but also as an "ethos of engagement", emphasising openness to socio-spatial possibilities and attentiveness to the fluid dynamics of urban formation. Revisiting assemblage in relation to materiality invites us to see matter not as passive but as possessing distinct affordances and intensities. This framework shifts inquiry from static definitions to relational processes—asking not simply what something is, but what it does and what it produces (Holubec, 2014; Yaneva, 2012).

By 'material paradox', I refer to the coexistence of ecologically symbolic materials (bamboo or banana leaves) with unsustainable practices and dependencies (plastic packaging and reliance on non-local organic goods). This paradox is not incidental but is produced and maintained through the entangled assemblage of governance, aesthetics, infrastructure, and economic adaptation at the market. This theoretical framework will guide the following analysis, from examining the market's initial branding to unpacking its daily material politics, moving us beyond simple binaries of success or failure towards a more inclusive and processual understanding of placemaking.

4 Methodology

This study employs an in-depth, qualitative case-study design to examine how sustainability is performed through materials. Barongan Market was selected purposively for three reasons. First, it is located in recent peri-urban desakota case that has sought to defend its local setting amid significant urban influx and development pressures, making it an information-rich site for interrogating the chapter's central 'material paradox'. Second, access and depth were feasible: an ongoing university–community collaboration enabled repeated entry to the site and understanding the engagement between both sides. Third, Barongan Market is locally influential: launched in late 2022, its curated "natural" aesthetic and bamboo-token payment system make it a deeply situated case reflecting complex social and material entanglements.

A qualitative, interpretive approach (Brinkmann, 2023; Creswell, 2010; Lucas, 2015), rooted in Southern urbanism and assemblage thinking, was adopted to surface lived material realities. Over 6 weeks, five complementary methods were combined to triangulate what people say and what materials do: participant observation, semi-structured interviews, focus group discussions, and informal conversations, alongside a review of visual and documentary materials. A micro-ethnographic method (Fetterman, 2009; Alvehus & Crevani, 2022; Powell, 2010) was also used to capture high-density interactions typical of temporal markets by staying in-and-off site for a period of time.

Approximately 40 participants were engaged through purposive sampling (Tongco, 2007), seeking variation by role (vendors, organisers, artisans, officials). Eligibility emphasised by their regular involvement in the market event and management. Recruitment of the participants was combined a direct approach and introductions by the university partner and the market's management. The interview was about 40–90 min depended on the participants' response, and it was recorded using a recorder to be transcribed later. While visitor perspectives were limited, the study focuses on those directly embedded in market practices and development, aligning with the aim to understand locally performed sustainability. Interviews explored themes such as material choices, aesthetics, and environmental narratives, while immersive direct observation positioned the researcher as a co-producer of the situated knowledge (Haraway, 1988). Data were analysed through iterative descriptive, process, and thematic coding (Saldana, 2009), across fieldnotes, transcripts, and visuals. Anchored in assemblage thinking, the analysis focused on relational dynamics—how materials acquire meaning and enact contested sustainability.

5 Findings

This section presents key findings from ethnographic fieldwork at Barongan Market, organised into two themes. The first explores the material paradox between its green branding and grounded realities. The second examines the active role of materials in shaping the market's socio-political dynamics. These findings reveal how sustainability in temporal placemaking is negotiated through everyday material practices.

5.1 Material Paradox: The Disjuncture Between Green Branding and Grounded Realities

Barongan Market is widely promoted as a paragon of eco-conscious placemaking, presented as a temporal space deeply connected to nature. This market is a product of speculation from below (Iqbal & Shafique, 2025), developed through a long-term university–community collaboration between Petra Christian University (PCU) in Surabaya and Desa Mojotrisno (Mojotrisno Village), in Jombang Regency, Indonesia.

University–community collaborations like this play a crucial role in shaping spatial interventions, acting as a catalyst for change through experiential pedagogy, service-learning, and research collaborations (Sara & Jones, 2018; Pak & De Smet, 2022; Forsyth et al., 2000). In Indonesia, such engagements also align with the broader mission of decolonising higher education and delivering direct societal impacts (Altbach, 2004; Suwignyo, 2024). This responsibility is formalised in the national Tri Dharma of Higher Educational regulation (UU, 1961), which mandates universities to conduct research, teaching, and community service activities.

Since 2021, Mojotrisno Village has collaborated with Petra Christian University mainly through an annual architectural service-learning module. The partnership began with a modest initiative involving local batik artisans in the village but quickly expanded to include broader programmes and more diverse stakeholders, including the local village government. The Barongan Market itself was officially launched in late 2022, supported by community service grants and research funding from the Indonesian Ministry of Higher Education. The initiative was led by Mr L from the interior department at PCU and drew inspiration from SEKALA, a bamboo communal building developed during their 2021 architecture service-learning cycle (Fig. 1).

Barongan Market is promoted as a nature-driven, temporal market rooted in traditional and sustainable practices, such as non-MSG foods, natural dyes, and biodegradable packaging. It reflects how nature is enacted differently in the Global South (Droz et al., 2022; Rossini, 2024), serving both environmental and economic goals. This branding aligns with broader village development agendas, helping Mojotrisno reimagine its identity from a former slum to a model of green innovation. In 2024, the village even hosted a National Environmental Day event in collaboration with

Fig. 1 SEKALA (Bamboo Communal Building) (*Source*: Author)

Fig. 2 Barongan Market during National Environmental Day Celebration 2024 (*Source*: Author)

the city-level Environmental Agency (Fig. 2). These public performances have successfully attracted attention from municipal and provincial authorities, positioning the market as a success story. As Mr L stated, "*Since the very beginning of our collaboration, we wanted to create something different and long-lasting.... showcasing these local potentials through an art and culinary market with environmental concerns*".

However, ethnographic observations reveal a stark material paradox: a disjuncture between this curated image and its material realities. A key example is the deteriorating bamboo used in stalls, signage, and the central SEKALA structure. While celebrated for its sustainability (Adebowale & Agumba, 2024), untreated

Fig. 3 SEKALA and its bamboo material problems (*Source*: Author)

bamboo has a short lifespan and is susceptible to damage (Hailemariam et al., 2023; Bui et al., 2017). Frequent repairs and struggles against decay undermine its image as a low-impact material (Fig. 3). A key issue is the limited local expertise in bamboo treatment, a gap left by an inadequate knowledge transfer process during the university collaboration. As Mr. S, a local craftsman, noted: "*We rushed to open the market, so some stalls were bought from outside. I personally can make them, but it takes time. However, I'm still the one fixing thing when it breaks*". His experience underscores how the prioritisation of instant aesthetics over developing local craftsmanship turns bamboo into a sustainable commodity to improve rural livelihood (Lobovikov et al., 2012) rather than a vehicle for building community capacity. Symbolic representations of sustainability often clash with ecological and logistical limitations (Lyon & Montgomery, 2015; Williams, 2024). Luckily, Mr S continues to assist in bamboo-related activities and workshops initiated by the university. However, his experience underscores a missed opportunity for sustainable knowledge transfer with the locals. The dependence on externally fabricated bamboo elements indicates a prioritisation of instant aesthetic appeal over the longer-term development of local material ecosystems and craftsmanship. The knowledge transfer envisioned in the design process failed to materialise. Here, bamboo functions as a sustainable commodity rather than a vehicle for knowledge transfer to residents who work with it on their daily basis.

Another significant contradiction is the re-emergence of single-use plastics, particularly during the expanded night-time market events. While the Sunday market more rigorously enforces a no-plastic policy, the night market reveals a different reality where plastic bags and containers are common. This shift reflects practical difficulties and ingrained habits (Linder et al., 2022). A food vendor, Mrs. N, commented: "*Changing mindsets and habits about using non-plastic materials is difficult to implement. Another problem is that leaf packaging depends on the weather. When it's too windy or rainy, it's hard to find leaves. Sometimes we use paper or cardboard, but we must purchase it anyway*". Even supposedly "natural" materials like banana leaves are often procured externally, undermining the ideal of local self-sufficiency.

This material paradox reveals the complexity of translating sustainability ideals into practice, especially in dynamic and resource-limited contexts (Bäckman, 2024; Håkansson et al., 2012). While the market has achieved public recognition and visibility, this success often hinges on surface-level performance—driven by aesthetic appeal, external expectations, and economic pressure—rather than a fully integrated ecological ethic and systemic changes.

Limited financial capacity for infrastructure maintenance, and ongoing logistical challenges all influence decisions that may compromise environmental principles. In this way, Barongan Market exemplifies how sustainability can become commodified (Cavanagh & Benjaminsen, 2017), where visual and narrative coherence overshadow material and systemic accountability. Yet, this contradiction is not a simple failure but a reflection of the market's complex assemblage of aspirations, constraints, and adaptations characteristic of Southern urbanism.

5.2 Material Agency and the Shaping of Social, Economic, and Political Assemblages

The materials at Barongan Market are far from passive components; they are active agents that shape and are shaped by the market's dynamic assemblage. These materials influence interactions and perceptions and are embedded in the market's contested sustainability. This lens shifts focus from what materials are to what they do—their roles in producing movements, relations, and tensions (Holubec, 2014; Yaneva, 2012). At Barongan, this means analysing the assemblage behind the material paradox and the relational forces that sustain it.

The pervasive presence of bamboo, for instance, extends far beyond its aesthetic contribution; it actively shapes spatial experience and the potential range of activities and social interactions within the market. Bamboo structures demarcate vendor stalls, informal seating zones, and the central SEKALA building. These spatial-material forms facilitate a wide range of activities—vendors selling goods, visitors resting and eating, and community members convening for events (Fig. 4). The flexible and temporary nature of bamboo seating, which can be easily rearranged, illustrates how the material enables adaptable spatial practices within a temporal context.

However, the inherent material limitations of untreated bamboo also actively constrain these possibilities in the long term. As evidenced by the frequent need for repairs and the eventual dismantling of structures like the bamboo stalls and their informal seating areas, the material reality of bamboo's degradation directly impacts the intended function, longevity, and even the symbolic perception of these spaces. The material condition of the bamboo infrastructure thus actively shapes the market's physical form, its capacity to host activities, and its overall stability as a temporal place.

Fig. 4 Barongan's flexible seating arrangement (*Source*: Author)

Mr S additionally mentioned, *"The university involved a bamboo artisan from Yogyakarta. But instead of transferring their knowledge, they ended up just working on the project. Even the students couldn't keep up. In the end, I helped complete the building. Sadly, no other villagers, especially the youth, were involved. But to be honest, bamboo was never really our concern, so only few people are interested*". This statement reveals how missed opportunities in knowledge transfer interact with material dependency. Bamboo, in this context, becomes an agent that organises labour, exposes institutional gaps, and reveals asymmetries between external collaborators and local communities within the assemblage.

Materials also shape the market's economic assemblage. The use of bamboo tokens as a form of currency is a striking example. This deliberate material choice reinforces the market's eco-cultural branding and creates an economic system distinct from conventional cash-based exchanges. Visitors convert rupiah into tokens at designated booths, and vendors then redeem them post-event. This not only structures specific economic roles within the market but also enhances its identity as a place-based, alternative economy—mirroring the model of Pasar Papringan in Yogyakarta (Widyastiti et al., 2022; Elfrida, 2021).

However, the financial challenges faced by the market, including the costs of operation and the fluctuating income derived from a percentage of vendor earnings (paid in bamboo tokens), demonstrate how the broader economic realities can strain the maintenance and viability of this material-based economic system. The material form of the currency is thus actively entangled with the market's economic health and the ongoing challenges of sustaining its unique operational model. Furthermore, the materials used in the food stalls and packaging actively contribute to shaping the market's economic and environmental identity.

The development of Barongan Market is not purely grassroots but shaped by village government influence, particularly the village head, Mr N. His strategic restructuring of the market's management created tensions with early contributors, such as the batik artisan and early youth group, who felt sidelined or withdrew. Mr. A, the local batik artisan mentioned, "*I think the role of local government should be in facilitating and supporting the initiative, it is hard if the grassroot initiative like Barongan should follow with the village agenda and managerial system. Therefore, I have decided to step down from the managerial role, even it is hard for me*". These shifts in governance affect decisions on materials and maintenance, with the market's declining physical conditions reflecting its insider political dynamics.

Material interventions themselves, such as the SEKALA and the village master plan developed through collaboration with university partners, also possess a form of political and social agency. The SEKALA building, selected for its distinctive design, became a focal point and a catalyst for subsequent development, demonstrating how a material object can actively initiate and direct further actions, investments, and initiatives within the village. The village master plan, although an intangible planning document, acts as a powerful material-like agent by providing a formal framework and justification for seeking external funding and directing the materialisation of future village development projects.

The involvement of external actors—such as a partnering university and government agencies—reveals the complex material dynamics at Barongan Market. Prefabricated bamboo stalls and externally sourced features like the playground illustrate how outside production influences local materiality. As Mr N, the village head, noted: "*We bought many items from nearby villages to maintain material consistency, though small items like bamboo stalls were made locally, often with Mr S.'s help*". These decisions reflect both material constraints and efforts to construct a cohesive market identity.

The concept of "ownership" and "belonging" within the community is also actively shaped by material realities and interventions. The initial enthusiasm and sense of collective ownership among the early initiators and vendors were tied to the physical manifestation of the market and its unique material identity. However, as political tensions arose and management shifted, the sense of ownership over the market's material assets became fragmented, contributing to issues of maintenance and care.

In conclusion, the materials at Barongan Market are far from inert or passive components of a static setting. They are active participants in the ongoing construction, negotiation, and contestation of the market as a place. The durability and maintenance challenges posed by bamboo, the fluctuating presence and implications of plastic, the symbolic and practical functions of the bamboo currency, and the ways in which material interventions are intertwined with political power struggles and economic realities all powerfully demonstrate the significant and often unpredictable agency of materials within this dynamic assemblage.

Understanding Barongan Market requires acknowledging how these diverse material agencies interact with human intentions, evolving social relationships, persistent economic pressures, and shifting political agendas, collectively creating a complex and frequently contradictory reality that challenges simplistic notions of sustainability and placemaking. The contested nature of sustainability at Barongan Market is thus not merely an abstract concept but is deeply embedded in the active, dynamic, and often unpredictable lives of the materials that physically and symbolically constitute the market.

6 Discussion and Beyond

By analysing Barongan Market through the lens of assemblage theory, this chapter has brought into sharp focus the complexities that characterise sustainability in the placemaking practices of the Global South. This discussion reorients temporal placemaking discourse towards an understanding of sustainability as a negotiated and assemblage materiality process, not merely a visual or discursive outcome. Read through an assemblage lens, the Barongan case suggests that sustainability in comparable peri-urban, eco-branded markets is provisional and continually negotiated rather than settled. In practice, it operates as a performance, an assemblage of aesthetics and external validation mechanisms that are often at odds with grounded lived realities.

At Barongan Market, the material paradox is stark. Bamboo and banana leaves function as potent symbols of eco-conscious design and careful development of local potential, which are beneficial for attracting visitors and branding the village. However, the material realities of decay and plastic use reveal the fragility of this narrative. These conditions reflect not just tradition but uneven knowledge transfer, economic limits, and market-driven pressures, which hinder deeper ecological transitions (Pereira et al., 2018; Scoones et al., 2020). Sustainability at Barongan Market thus functions more as a curated identity than lived reality, aligning with what Fairhead et al. (2012) term "green grabbing" to satisfy external expectations.

The persistence of these contradictions is best understood through the lens of assemblage theory, which reconfigures our approach to space, place, and materiality. Rather than viewing materials as passive components of a pre-existing framework, assemblage thinking—as developed by Deleuze and Guattari (1987) and elaborated by McFarlane and Anderson (2011)—sees space as a constellation of heterogeneous relations. Bamboo is not simply sustainable or unsustainable; its role shifts depending on treatment knowledge, weather conditions, cultural valuation, and governance structures.

The sustainability of Barongan Market is further entangled with local politics. As the market transitions from a community-driven initiative to a semi-institutionalised village asset governed by local village government, power struggles emerge over representation, management, and ownership. The sidelining of early collaborators underscores how placemaking is not only spatial but can be political (Salzman & Yerace, 2018; Pierce et al., 2011). The symbolic power of the bamboo market extends into economic and spatial governance.

The implications for architectural and planning research are clear. First, design practices must move beyond eco-aesthetics and invest in slow sustainability, which prioritises repair, knowledge co-production, and the cultivation of local material ecosystems. This entails supporting community infrastructures for training, maintenance, and feedback rather than one-time installations. Sustainable craftsmanship needs to be cultivated rather than just relying on external resources and support. It also includes conscious care of local material production. Second, research must attend to material politics—not just what materials are used, but who controls them, who maintains them, and how they operate within networks of power.

Finally, this research reaffirms that sustainability cannot be disentangled from social justice, labour, and epistemic plurality. The future of placemaking in the Global South depends not on perfect materials or ideal images, but on relational accountability—a commitment to understanding how people and materials co-constitute places over time. The contradictions of Barongan Market are not failures but symptoms of deeper structural tensions in how sustainability is practiced, imagined, and institutionalised.

In closing, Barongan Market offers a compelling lens to re-examine the architecture of sustainability itself—not as a design output but as an emergent, uneven, and deeply political material process. It is within this tension—between the image and the infrastructure, the symbolic and the grounded—that the future of equitable and durable placemaking must be envisioned.

Acknowledgements The author would like to thank the village government and residents of Mojotrisno Village, as well as the Architecture Department of Petra Christian University (PCU), for their kindness and assistance during the doctoral fieldwork. The publication of this article was also supported by the Lembaga Pengelola Dana Pendidikan (LPDP) of the Republic of Indonesia, whose scholarship made this work possible.

References

Adebowale, O. J., & Agumba, J. N. (2024). Bamboo in sustainable construction: Effects on productivity and safety. *International Journal of Productivity and Performance Management, 74*(11), 1–20.

Altbach, P. G. (2004). In P. G. Altbach & T. Umakoshi (Eds.), *Asian universities: Historical perspectives contemporary challenges*. Johns Hopkins University Press.

Alvehus, J., & Crevani, L. (2022). Micro-ethnography: Towards an approach for attending the multimodality of leadership. *Journal of Change Management, 22*(3), 231–251.

Anderson, R. S., Bruun, O., & Kalland, A. (1996). Asian perceptions of nature: A critical approach. *Pacific Affairs, 69*(4), 563.

Bäckman, M. (2024). (Un)sustainable everyday practices sociomateriality shaping sustainability in an urban district. *Journal of Consumer Culture, 24*(1), 100–119.

Beasy, K., & Corbett, M. (2021). What counts as sustainability?: A sociospatial analysis. *Environmental Sociology, 7*(4), 327–337.

Brinkmann, S. (2023). *Qualitative interviewing*. Oxford University Press.

Buchanan, I. (2017). Assemblage theory, or, the future of an illusion. *Deleuze Studies, 11*(3), 457–474.

Bui, Q.-B., Grillet, A.-C., & Tran, H.-D. (2017). A bamboo treatment procedure: Effects on the durability and mechanical performance. *Sustainability, 9*(9), 1444.

Burns, T. R., & Witoszek, N. (2012). Brundtland report revisited: Toward a new humanist agenda. *Journal of Human Ecology (Delhi, India), 39*(2), 155–170.

Caprotti, F., & Cowley, R. (2017). Interrogating urban experiments. *Urban Geography, 38*(9), 1441–1450.

Cavanagh, C. J., & Benjaminsen, T. A. (2017). Political ecology, variegated green economies, and the foreclosure of alternative sustainabilities. *Journal of Political Ecology, 24*(1), 200.

Creswell, J. (2010). Research design: Qualitative, quantitative, and mixed methods approaches. March.

Deleuze, G., & Guattari, F. (1987). *A thousand plateaus: Capitalism and schizophrenia* (Vol. 19). University of Minnesota Press.

Dovey, K., Cook, B., & Achmadi, A. (2019). Contested riverscapes in Jakarta: Flooding, forced eviction and urban image. *Space and Polity, 23*(3), 265–282.

Droz, L., Brugnach, M., & Pascual, U. (2023). Multilingualism for pluralising knowledge and decision making about people and nature relationships. *People and Nature (Hoboken, N.J.), 5*(3), 874–884.

Droz, L., Chen, H.-M., Chu, H.-T., Fajrini, R., Imbong, J., Jannel, R., Komatsubara, O., et al. (2022). Exploring the diversity of conceptualizations of nature in East and South-East Asia. Humanities & Social Sciences. *Communications, 9*(1). https://doi.org/10.1057/s41599-022-01186-5

Droz, L., Fricke, M. F., Heroor, N., Jannel, R., Komatsubara, O., Concordia, M. A., Lagasca-Hiloma, P. M., Matangcas, J. J., & Sihombing, H. H. (2024). Environmental philosophy in Asia: Between eco-orientalism and ecological nationalisms. *Environmental Values, September.* https://doi.org/10.1177/09632719241276999

Ducarme, F., & Couvet, D. (2020). What does 'nature' mean? *Palgrave Communications, 6*(1), 1–8.

Elfrida, T. (2021). Food tourism and sustainable development model: Case study of Pasar Papringan. *Journal of Tourism Destination and Attraction, 9*(1), 79–90.

Fairhead, J., Leach, M., & Scoones, I. (2012). Green grabbing: A new appropriation of nature? *The Journal of Peasant Studies, 39*(2), 237–261.

Fetterman, D. M. 2009. *Ethnography: Step-by-step* (3rd ed.). Applied social research methods. SAGE.

Forsyth, A., Henry, L., & McGirr, P. (2000). Service learning in an urban context: Implications for planning and design education. *Journal of Architectural and Planning Research, 17*(3), 236–259.

Glazener, A., Wylie, J., van Waas, W., & Khreis, H. (2022). The impacts of car-free days and events on the environment and human health. *Current Environmental Health Reports, 9*(2), 165–182.

Griep, Y., Kraak, J. M., & Beekman, E. M. (2023). Sustainability is dead, long live sustainability! Paving the way to include 'the people' in sustainability. *Group & Organization Management, 48*(3), 966–980.

Hailemariam, E. K., Hailemariam, L. M., Amede, E. A., & Nuramo, D. A. (2023). Identification of barriers, benefits and opportunities of using bamboo materials for structural purposes. *Engineering Construction and Architectural Management, 30*(7), 2716–2738.

Håkansson, M., Leshed, G., Blevis, E., Nathan, L., & Mann, S. (2012). Simple, sustainable living. In *CHI '12 Extended abstracts on human factors in computing systems*. ACM. https://doi.org/10.1145/2212776.2212723

Hakkarainen, V., Anderson, C. B., Eriksson, M., van Riper, C. J., Horcea-Milcu, A., & Raymond, C. M. (2020). Grounding IPBES experts' views on the multiple values of nature in epistemology, knowledge and collaborative science. *Environmental Science & Policy, 105*(March), 11–18.

Haraway, D. (1988). Situated knowledges: The science question in feminism and the privilege of partial perspective. *Feminist Studies: FS, 14*(3), 575.

Holubec, P. (2014). Assemblage thinking in urban studies: How to conceive of a city? *Advanced Engineering Forum, 12*(November), 17–22.

Iqbal, M. N. M., & Shafique, T. (2025). Speculation from below: How temporal market sustains local heritage in Indonesia's *desakota* region. *International Planning Studies, 30*(1–2), 141–155.

Keeble, B. R. (1988). The Brundtland report: 'Our Common Future'. *Medicine and War, 4*(1), 17–25.

Kuhlman, T., & Farrington, J. (2010). What is sustainability? *Sustainability, 2*(11), 3436–3448.
Latour, B. (2005). *Reassembling the social: An introduction to actor-network-theory*. Clarendon Lectures in Management Studies. Oxford University Press.
Linder, N., Giusti, M., Samuelsson, K., & Barthel, S. (2022). Pro-environmental habits: An under-explored research agenda in sustainability science. *Ambio, 51*(3), 546–556.
Lobovikov, M., Schoene, D., & Yping, L. (2012). Bamboo in climate change and rural livelihoods. *Mitigation and Adaptation Strategies for Global Change, 17*(3), 261–276.
Lucas, R. (2015). *Research methods for architecture*. Laurence King Publishing.
Lyon, T. P., & Wren Montgomery, A. (2015). The means and end of greenwash. *Organization & Environment, 28*(2), 223–249.
Mady, C., Khan, S. S., Ortiz, C., Stephan, J., Makakavhule, K., Kamal, O., & Meza, M. (2025). Southern urbanism as a negotiation of past, present, and future. *International Planning Studies, 30*(1–2), 1–7.
McFarlane, C., & Anderson, B. (2011). Thinking with assemblage: Thinking with assemblage. *Area, 43*(2), 162–164.
McGee, T. G. (1989). Urbanisasi or Kotadesasi? Evolving patterns of urbanization in Asia. *Urbanization in Asia*, 93–108.
Nail, T. (2017). What is an assemblage? *SubStance, 46*(1), 21–37.
Nielsen, L. W. (2004). The 'nature' of 'nature': The concept of nature and its complexity in a western cultural and ethical context. *Global Bioethics, 17*(1), 31–38.
Pak, B., & De Smet, A. (Eds.). (2022). *Experiential learning in architectural education: Design-build and live projects*. Routledge Focus on Design Pedagogy. Routledge.
Pereira, L., Karpouzoglou, T., Frantzeskaki, N., & Olsson, P. (2018). Designing transformative spaces for sustainability in social-ecological systems. *Ecology and Society, 23*(4). https://doi.org/10.5751/ES-10607-230432
Pierce, J., Martin, D. G., & Murphy, J. T. (2011). Relational place-making: The networked politics of place. *Transactions (Institute of British Geographers: 1965), 36*(1), 54–70.
Plumwood, V. (2019). Feminism and the mastery of nature. In *Ideals and ideologies* (11th ed., pp. 487–493). Routledge, 2019. | "Tenth edition, published by Routledge, 2017"-T.p. verso.: Routledge.
Powell, K. (2010). Viewing places: Students as visual ethnographers. *Art Education, 63*(6), 44–53.
Rajendran, L. P., Raúl, L., Chen, M., Andrade, J. C. G., Akhtar, R., Mngumi, L. E., Chander, S., Srinivas, S., & Roy, M. R. (2024). The 'peri-urban turn': A systems thinking approach for a paradigm shift in reconceptualising urban-rural futures in the global south. *Habitat International, 146*(103041), 103041.
Rossini, F. (2024). Urban design and informal settlements: Placemaking activities and temporary architectural interventions in BaSECo compound. *Urban Design International, 29*(1), 4–24.
Salama, A. M., Patil, M. P., Elsemellawy, A. N., Abudib, H. H., Almansor, N. A., MacLean, L., & van Riel, K. (2024). People–place narratives as knowledge typologies for social sustainability: Cases from urban contexts in the global south. *Buildings, 14*(4), 1001.
Saldana, J. (2009). *The coding manual for qualitative researchers*. SAGE Publications.
Salzman, R., & Yerace, M. (2018). Toward understanding creative placemaking in a socio-political context. *City Culture and Society, 13*(June), 57–63.
Sara, R., & Jones, M. (2018). The university as agent of change in the city: Co-creation of live community architecture. *International Journal of Architectural Research Archnet-IJAR, 12*(1), 326.
Schindler, S. (2017). Towards a paradigm of southern urbanism. *City, 21*(1), 47–64.
Scoones, I., Stirling, A., Abrol, D., Atela, J., Charli-Joseph, L., Eakin, H., Ely, A., et al. (2020). Transformations to sustainability: Combining structural, systemic and enabling approaches. *Current Opinion in Environmental Sustainability, 42*(February), 65–75.
Shafique, T. (2022). Re-thinking housing through assemblages: Lessons from a Deleuzean visit to an informal settlement in Dhaka. *Housing Studies, 37*(6), 1015–1034.
Shin, H. B. (2021). Theorising from where? Reflections on de-centring global (southern) urbanism. In *Global urbanism* (1st ed., pp. 62–70). Routledge.

Sui, D. Z., & Zeng, H. (2001). Modeling the dynamics of landscape structure in Asia's emerging Desakota regions: A case study in Shenzhen. *Landscape and Urban Planning, 53*(1-4), 37–52.

Suwignyo, A. (2024). Higher education as an instrument of decolonisation: The community service programme in Indonesia, 1950–1969. *Asian Studies Review, 48*(3), 447–466.

Tongco, M. D. C. (2007). Purposive sampling as a tool for informant selection. *Ethnobotany Research and Applications, 5*(December), 147.

UU. (1961). *Undang-Undang Republik Indonesia Nomor 22 Tahun 1961 Tentang Pendidikan Tinggi (National Regulation of the Republic of Indonesia Number 2 Year 1961 on the Higher Education).* Indonesian Government.

Widyastiti, S. A., Alsyifa, E. T., Sayekti, R. B., & Pambuko, Z. B. (2022). The marketing strategy and benefits analysis of Papringan market in Indonesia: A narrative review of literature. *Urecol Journal. Part H: Social, Art, and Humanities, 2*(2), 44–48.

Williams, J. (2024). Greenwashing: Appearance, illusion and the future of 'green' capitalism. *Geography Compass, 18*(1), e12736.

Wood, A. (2020). Decolonising cities of the global south in the classroom and beyond. *The Town Planning Review, 91*(5), 535–552.

Yaneva, A. (2012). *Mapping controversies in architecture.* Ashgate Publishing.

Yan, W., Liao, J., & Zhai, H. (2024). Place-making research: A bibliometric, visualization, and thematic analysis. *Buildings, 14*(9), 2855.

Yudistira, F. (2017). Creative strategy for creating public space for creative tourism (case: Pasar Kaget at Sunda Kelapa Mosque). In *Proceedings of the 6th International Conference of Arte-Polis* (pp. 41–50). Springer Singapore.

Biophilic Urban Streetscapes: GIS-Based Approaches for Nature-Driven Placemaking in Uzbekistan

Jamshid Sodikov

Abstract Nature-driven placemaking, underpinned by biophilic design, provides transformative potential for urban environments in arid or dry-continental regions. In Sect. 2, "Community-Level Interventions", we demonstrate how Geographic Information Systems (GIS) can guide the strategic planning of nature-oriented streetscapes in Tashkent, Uzbekistan, by combining multi-temporal remote sensing with structured stakeholder engagement. First, we apply a 5-year Normalized Difference Vegetation Index (NDVI) analysis (2020–2024) to classify urban vegetation into four cover classes and compute district-level proportions of bare, sparse, and vegetated areas. These maps reveal that existing biophilic walking routes closely align with high-NDVI corridors along the canal network, while several districts show declines in sparse vegetation, pinpointing priority zones for greening. A pixel-wise ΔNDVI map further identifies localized "hotspots" of loss and gain.

Building on these empirical insights, we present an integrative, reproducible workflow—illustrated in a detailed process flow (Fig. 4)—that merges GIS analytics with participatory design. Community validation through Mahalla Committee workshops and professional charrettes confirms the accuracy of our corridor detections and refines on-the-ground planting proposals. Policy integration steps, in collaboration with the Municipal Planning Department, translate our statistical findings into updated GIS layers and budget targets for district greening quotas.

We conclude by offering a replicable roadmap for scaling this biophilic framework across Central Asian cities (e.g. Samarkand, Bukhara, Namangan), recommending the establishment of a regional Green Cities Consortium, embedding biophilic metrics into city master plans, and mobilizing community data streams for continual route optimization. This chapter's novel contribution is a concise, empirically validated model—grounded in both "big-data" mapping and local knowledge—for designing resilient, culturally resonant biophilic streetscapes in extreme-climate contexts.

J. Sodikov (✉)
Tashkent State Transport University, Tashkent, Uzbekistan
e-mail: sodiqov_j@tstu.uz

A. Cheshmehzangi et al. (eds.), *Designing Nature into Cities*, Urban Sustainability, https://doi.org/10.1007/978-981-95-9382-8_4

Keywords Biophilic design · Nature-driven placemaking · Urban streetscapes · GIS workflow · Participatory mapping · Climate adaptation

1 Introduction

Rapid urbanization across the world's dry-continental belt brings not only expanding concrete footprints but also a serious threat to public health: soaring summer temperatures, worsening air quality, and diminished access to restorative green space combine to increase heat-related illness, respiratory problems, and stress. In Tashkent, mean summer air temperatures regularly exceed 27 °C and Landsat-derived surface temperatures can top 50 °C on unshaded pavements during July heatwaves. Surveys of local residents and lively social-media debates confirm widespread community demand for more street trees, pocket parks, and cooler pedestrian routes. In this context, biophilic design—the intentional integration of nature into streets, façades, and public places—emerges as a powerful strategy to advance climate adaptation, improve air quality, support mental well-being, and strengthen place identity. Current scholarship frames "biophilic streets" as the gateway to a broader, nature-centred urbanism, delivering measurable gains in thermal comfort, biodiversity, and social vitality when vegetation is woven directly into the city fabric (Ryan et al., 2014).

1.1 Rationale

Uzbekistan exemplifies the intertwined challenges and opportunities of arid-zone cities. In Tashkent, impervious surfaces already cover roughly 78% of the urban area, while green spaces amount to only 5.6%—conditions that intensify the urban heat island to 1.5–1.6 °C above surrounding rural land. A recent Google Earth Engine analysis for the summers of 2020–2024 (Sharipov & Khayitmurodov, 2024) confirms persistent "hot corridors" along major arterial streets where vegetation falls below 10% and daytime land-surface temperatures climb 5–7 °C higher than on tree-lined boulevards of comparable width (see Sect. 3).

National policy is beginning to respond: Uzbekistan's *Strategy for the Transition to a Green Economy 2019–2030*, updated by Presidential Resolution No. 436 in December 2022, elevates urban greening and heat mitigation as priority adaptation measures but stops short of detailed guidance for community-scale implementation (UNDP, 2023). Municipal tree-planting campaigns often falter when species selection and planting sites are ill-suited to the hot-dry climate, resulting in high mortality and wasted resources. These gaps underscore the urgent need for an evidence-based, spatially explicit framework—grounded in local cultural values and ecological resilience—that can guide planners, policymakers, and communities in creating streetscapes that truly thrive under extreme climatic stress.

1.2 Problem Statement

Despite growing recognition of nature-based solutions, most Uzbek streetscapes remain dominated by wide asphalt carriageways, reflective concrete, and sparse ornamental planting. Without strategic intervention, climate models project that heatwave frequency in Central Asia will increase by 20–30% by mid-century under a moderate-emissions scenario (IPCC AR6 WG I, 2021), and ongoing desertification will exacerbate these extremes—together threatening pedestrian comfort, driving up cooling-energy demand, and eroding urban biodiversity. The core problem addressed in this chapter is therefore how to translate the principles of biophilic design into practical, GIS-supported actions that retrofit or newly build streetscapes capable of moderating microclimate, reinforcing ecological connectivity, and nurturing socio-cultural identity under arid-continental conditions.

1.3 Objectives

This chapter has four interlinked objectives. First, it quantifies urban heat-island exposure, maps vegetation distribution, and identifies shade deficits along representative streets in Uzbek cities by leveraging multi-source GIS and remote-sensing datasets—including Landsat-8, Sentinel-2, and cadastral layers. Second, it develops an integrative framework that melds spatial analytics with participatory design workflows, guiding local stakeholders step by step from diagnostic mapping to the co-creation of biophilic interventions. Third, it demonstrates through community-level case studies how strategically placed green corridors, drought-tolerant tree canopies, and culturally resonant streetscape elements can lower surface temperatures and improve walkability. Finally, it distils a set of policy-oriented recommendations—covering governance structures, species selection criteria, watering regimes, and monitoring protocols—that support long-term climate resilience and place-making goals in arid-continental urban contexts.

1.4 Chapter Overview

Section 2 reviews the theoretical underpinnings of biophilic urbanism and synthesizes global best practice relevant to arid climates. Section 3 details the GIS and GEE workflow used to map thermal comfort, canopy cover, and pedestrian exposure in Tashkent. Section 4 presents three community-scale design scenarios, evaluating their thermal-performance and socio-ecological co-benefits. Section 5 translates these insights into a policy roadmap, highlighting financing models, maintenance strategies, and metrics for success. The conclusion (Sect. 6) reflects on the replicability of the proposed framework across Central Asia and outlines future research directions.

By grounding cutting-edge spatial analysis in the socio-cultural fabric of Uzbekistan, this chapter offers a replicable blueprint for nature-driven placemaking that can turn heat-stressed streets into verdant, people-centred corridors—demonstrating that biophilic design is not a luxury, but a climatic necessity in the twenty-first-century Silk Road cities.

2 Background Knowledge

This chapter draws on two intertwined bodies of knowledge. The first is the theory of biophilia and its translation into biophilic design, which provides the foundational rationale for nature-driven placemaking. The second is the suite of GIS-supported methods from landscape ecology and urban climatology, which supply the tools for spatial analysis and mapping. Together, these conceptual and methodological streams form a scaffolding that enables us to plan, evaluate, and refine biophilic streetscapes in Tashkent's extreme-climate context.

2.1 The Biophilia Concept and Its Architectural Offspring

The biophilia hypothesis, first articulated by Wilson (1984), posits that humans possess an innate tendency to affiliate with living systems and natural processes. This evolutionary bond, Wilson argued, has profound implications for human well-being and has been elaborated by Kellert and Wilson (1993) into nine distinct value dimensions—ranging from the utilitarian appreciation of natural resources to the aesthetic and emotional benefits derived from encountering living forms.

Building on this theoretical foundation, Stephen R. Kellert (2008) developed a three-part design framework for biophilic architecture. He categorized biophilic design elements into "Nature in the Space", which refers to direct experiences of vegetation, water, and other natural features; "Natural Analogues", encompassing materials and patterns that evoke the qualities of the natural world; and "Nature of the Space", which describes spatial configurations—such as prospect and refuge—that resonate with our innate preferences for certain environmental conditions. Kellert catalogued more than 70 attributes across these categories, offering designers a rich palette of strategies for embedding life-affirming qualities into the built environment.

In parallel, Terrapin Bright Green's influential report, *14 Patterns of Biophilic Design* (2014), synthesized empirical evidence linking specific design patterns—such as visual connection with nature, dynamic and diffuse natural lighting, presence of water, material connection to natural elements, mystery, and risk/peril—to measurable cognitive, psychological, and physiological outcomes. By translating abstract theory into a "pattern language", this work has equipped practitioners with a clear, evidence-based roadmap for creating health-supportive environments.

Although these strands of research converge on the shared goal of restoring directed attention, lowering cortisol levels, and reinforcing place identity, their practical application must be tailored to local climatic and cultural conditions. In arid-continental cities like Tashkent, lush green walls and dense understorey plantings may be neither feasible nor water-wise. Instead, designers can draw on the same underlying biophilic patterns—shade-casting tree canopies, evaporative water features, colour-rich xerophytic vegetation, and street geometries that promote natural ventilation—to deliver indoor-outdoor comfort, ecological connectivity, and socio-cultural resonance within a hot-dry environment.

2.2 Key Theoretical Lenses for Nature-Driven Streetscapes

Lens	Core idea	Relevance to Uzbek streets	Foundational sources
Prospect–refuge theory	Humans seek open vistas for surveillance coupled with enclosed refuges for safety.	Pedestrian-level sightlines along axial boulevards balanced by shaded alcoves and recessed shopfronts.	Appleton (1975); Kellert (2008)14-Patterns-of-Biophili…95
Attention restoration theory (ART)	Soft-fascinating settings replenish cognitive capacity depleted by directed attention.	Pocket gardens, patterned paving, and dappled light under *Ulmus pumila.*	Kaplan & Kaplan (1989); Ulrich (1991)
Landscape ecology	Spatial configuration—patch, corridor, matrix—governs ecological flows.	GIS-mapped green corridors stitching fragmentary mahalla courtyards and canal banks.	Forman & Godron (1986); Dramstad et al. (1996)
Ecosystem-service urbanism	Streetscapes supply regulating (shade, cooling), supporting (habitat), and cultural (heritage) services.	Native *Juniperus seravschanica* rows lower PET[a] by 4–6 °C during July heatwaves.	Tzoulas et al. (2007); Gómez-Baggethun et al. (2013)

[a]Physiological equivalent temperature

2.3 GIS and Remote-Sensing Foundations

Geographic Information Systems make it possible to layer environmental, infrastructural, and socio-cultural data into a single analytic framework. In this chapter, we rely on four core GIS-based methods to diagnose and design biophilic streetscapes in Tashkent. First, we perform thermal mapping by combining Landsat

8 Collection 2 land-surface temperature rasters (30 m resolution) with Sentinel-2 Normalized Difference Vegetation Index (NDVI) composites. This fusion reveals persistent “hot corridors”—sections of roadway whose mean surface temperatures exceed those of adjacent, better-vegetated blocks by more than 5 °C—a pattern that our Google Earth Engine time-series (2020–2024) confirms year after year.

Second, we carry out shade and solar-access analysis using LiDAR-derived building height models. By simulating hourly sun and shadow dynamics and overlaying these results onto a pedestrian-network centrality map, we identify the street segments where new tree canopies will deliver the greatest increase in shade for walkers during peak heat.

Third, we map least-cost ecological networks with tools from circuit theory and graph analysis. In practice, this involves treating existing green patches—such as riparian fragments along the Ankhor Canal and the Botanic Garden—as “nodes” in a resistance surface, then computing the optimal alignments for street-tree plantings that would reconnect these nodes and restore habitat continuity.

Finally, we integrate participatory GIS to ground our analysis in local knowledge. We enable mahalla residents to geo-tag their lived experiences—pinpointing heat-stress hotspots, reporting malfunctioning fountains, or highlighting culturally significant trees. These community inputs form an essential feedback loop, ensuring that our spatially explicit recommendations resonate with local needs and values.

2.4 Fundamental Definitions

Term	Working definition for this chapter
Biophilic urban streetscape	A linear public realm intentionally structured to foster positive human–nature interactions while delivering microclimatic, ecological, and cultural benefits.
Nature-driven placemaking	A co-creative process that uses empirical environmental data and local cultural narratives to embed living systems and their analogues into everyday settings.
Green corridor (urban)	A GIS-identified, vegetation-dominated pathway that supports thermal comfort, active mobility, and species movement across the street network.
Arid-continental climate design logics	Strategies that maximize shade, evapotranspirative cooling, albedo modulation and drought-resilient biodiversity under large diurnal and seasonal temperature swings.

2.5 Contextual Frame: Uzbekistan

Uzbek cities exhibit Soviet-era wide arterials, sparse ornamental planting, and rising motorization, amplifying Urban Heat Island intensity (1.5–1.6 °C). Concurrently, the Strategy for the Transition to a Green Economy 2019–2030 mandates a 30% rise in urban green cover, creating fertile ground for biophilic retrofits. Desertification pressure from the Aralkum and water scarcity underline the importance of native, low-water tree palettes (e.g. *Populus pruinosa*, *Elaeagnus angustifolia*) and GIS-guided irrigation zoning.

2.5.1 Synthesis

The literature affirms that biophilia provides the motivational grammar, while GIS affords the syntactic precision to script streetscapes that are climatically resilient and experientially rich. The subsequent sections translate these foundations into a step-by-step methodology and case-study evidence for Uzbekistan's community-level interventions.

3 Methodology

This section documents the end-to-end workflow used to quantify Urban-Heat-Island (UHI) intensity and vegetation deficits for every administrative district (*tuman*) in Tashkent over five consecutive summers (2020–2024). The procedure is implemented in Google Earth Engine (GEE) and summarized schematically in Fig. 1. Because municipal greening budgets, maintenance staff, and monitoring targets in Uzbekistan are organized at the district level, all statistics are aggregated to that spatial grain rather than to individual streets.

3.1 Spatial Frame

- *Primary units of analysis*—The 12 cadastral districts of Tashkent, provided by the State Cadastre Agency and stored in GEE as the Tash_2021 layer.
- *Region of interest (ROI)*—The dissolved union of the 12 districts (≈350 km^2). A 500 m outward buffer is applied during raster clipping to minimize edge effects when calculating neighbourhood statistics.

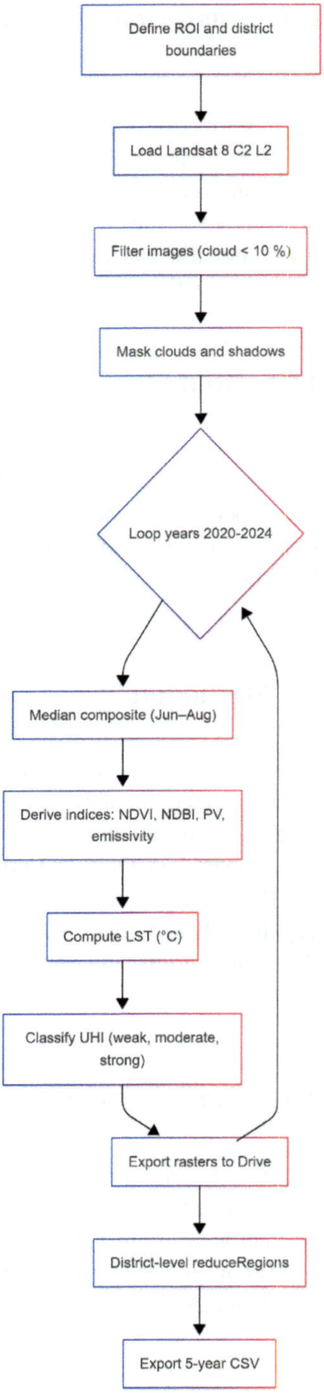

Fig. 1 GEE processing flowchart for district-level UHI assessment

3.2 Input Datasets

Thematic role	Dataset (GEE ID)	Native resolution	Time span used	Notes
Surface temperature and reflectance	Landsat 8 Collection 2, Level-2 SR & ST (LANDSAT/LC08/C02/T1_L2)	30 m	Jun 2020–Aug 2024	Surface-reflectance bands, brightness-temperature band, QA_PIXEL mask
Vegetation sharpening	Sentinel-2 MSI Level-2A (COPERNICUS/S2_SR)	10 m	Jun 2020–Aug 2024	Red and NIR bands used only for cross-checks; not aggregated
Validation	Uzhydromet automatic-weather stations	point	Jul 2020–Aug 2024	Three stations inside the ROI; 10-min air-temperature series

All imagery is re-projected to *WGS 84/UTM 42 N* (EPSG: 32642) and analysed at 30 m to match Landsat's nominal resolution.

3.3 Processing Pipeline (See Fig. 1)

Our workflow unfolds in four sequential stages: data preparation, annual composite generation, district-level aggregation, and time-series export.

3.3.1 Data Preparation

We begin by loading the Tash_2021 FeatureCollection into Earth Engine. The individual district polygons are preserved for later zonal statistics, while a dissolved union of all features defines the study area (region of interest, ROI) for raster clipping and export.

3.3.2 Annual Composite Generation

We filter the Landsat 8 Collection 2 Tier 1 Level-2 imagery to the June 1–August 31 window of each year, retaining only scenes with less than 10% cloud cover. A cloud-masking function examines the QA_PIXEL band and zeros out bits 3 (shadow) and 5 (cloud), ensuring that only clear-sky pixels contribute to our composites. For each year from 2020 to 2024, we then compute a median composite of all cloud-masked scenes within the ROI; this step both reduces residual clouds and minimizes phenological noise.

Next, we derive four spectral and biophysical indices from the summer composite. The Normalized Difference Vegetation Index (NDVI) and the Normalized

Difference Built-up Index (NDBI) use the Landsat shortwave infrared (SWIR) and near-infrared (NIR) bands, while the proportion of vegetation (PV) is calculated by linearly scaling NDVI between 0.2 and 0.5 and clamping to [0, 1]. Broad-band surface emissivity (ε) then follows as $\varepsilon = 0.986 + 0.004 \times PV$.

To obtain land-surface temperature (LST) in °C, we first convert the surface temperature band (ST_B10) to brightness temperature in Kelvin using the USGS gain and offset. We then apply the single-channel atmospheric correction formula to derive LST in Kelvin before subtracting 273.15.

We classify the resulting LST_C raster into three urban-heat-island (UHI) classes—weak, moderate, and strong—using city-wide thresholds at mean + σ and mean + 2σ. A 3×3 focal-mode filter is applied to the "strong" class to remove isolated pixels and produce a clean UHI layer. Finally, for each year, we export seven GeoTIFF layers (NDVI, NDBI, ε, LST_C, UHI_Weak, UHI_Moderate, and UHI_Strong) at 30 m resolution (EPSG:4326) to Google Drive, appending the year to each filename.

3.3.3 District-Level Aggregation

After all five annual composites have been produced, we calculate zonal statistics for each district polygon. Using a chained reducer, we extract the mean, minimum, maximum, median (50th percentile), and standard deviation for LST_C, NDVI, NDBI, and ε. These per-district summary metrics are combined into a single FeatureCollection containing 60 features (12 districts $\times$ 5 years) and 26 attribute fields.

3.3.4 Time-Series Export

The final step exports this aggregated table as a CSV file (UHI_Timeseries_2020_2024.csv), which can be loaded into external statistical software for longitudinal analysis and cross-district comparisons.

3.4 *Validation*

LST values extracted from 30 m buffers around the three AWS locations were paired with AWS air-temperature readings taken within ±5 min of Landsat over-pass time (≈10:25 UTC, 15:25 local). Linear regression yields MAE = 1.7 °C and $R^2 = 0.71$, consistent with published mid-latitude validations of the single-channel algorithm.

3.5 Justification for District-Scale Aggregation

Aggregating our indices at the district (tuman) level—rather than attempting to work at the granularity of individual street segments—rests on four interrelated considerations: policy alignment, data completeness, ecological coherence, and equity in analysis.

First, from a policy-leverage perspective, municipal budgets and greening mandates in Uzbekistan are allocated by district rather than by street. While street-level metrics offer fine-grained insight, they rarely map directly onto administrative budget lines or maintenance jurisdictions. By summarizing key metrics (e.g. mean NDVI, LST) across each district, our results feed naturally into existing planning and resource-allocation frameworks, thereby enhancing the likelihood that our recommendations can be adopted and funded.

Second, with respect to data completeness, the official street-centerline and metadata layers are often patchy in outlying mahallas, leading to gaps or biases if analyses rely solely on vector street networks. In contrast, cadastral district polygons are maintained as authoritative boundaries and benefit from rigorous national surveying protocols. District-level aggregation thus avoids the spurious "edge effects" and missing data that would undermine a street-segment analysis, ensuring more robust and reproducible statistics.

Third, *ecological coherence* favours district aggregation where "green infrastructure" extends beyond the confines of single corridors. Canopy corridors, courtyard clusters, canal belts, and pocket parks form interconnected mosaics that span multiple streets and blocks. Summarizing across a district captures these broader spatial patterns of vegetation and impervious cover, reflecting the true scale at which ecological processes (shade connectivity, stormwater attenuation, habitat provision) operate in an urban context.

Finally, from an *equity-analysis* standpoint, comparing dozens or hundreds of individual streets can obscure systemic disparities. District-level summaries enable straightforward side-by-side comparison across the 12 administrative units of Tashkent, highlighting which districts are relatively under-served by green cover or exposed to higher surface temperatures. This scale of analysis supports priority ranking and resource targeting, ensuring that interventions can be directed to the districts most in need of greening.

By anchoring our workflow in district-scale aggregation, we therefore align technical rigour with administrative practice, data availability, ecological logic, and equity objectives—laying the foundation for actionable, city-wide biophilic interventions.

3.6 Software Environment and Reproducibility

All of the earth-observation processing was carried out within the Google Earth Engine (GEE) platform; the complete GEE script (https://code.earthengine.google.com/?scriptPath=users%2Fosmijam%2Ftest%3AUHI_Multi_Year_Analysis) is publicly archived. To produce publication-quality choropleth maps, bivariate plots, and illustrative figures, we used QGIS 3.34 "Prizren". Finally, all statistical summaries, regression analyses, and kernel-density estimations described in Sect. 4 were implemented in a Python 3.10 environment (GeoPandas, SciPy), ensuring that every step from raw TIFF to final chart can be rerun by other researchers.

3.7 Limitations

Several caveats deserve emphasis. First, the 30 m spatial resolution blends narrow alleys, courtyard gaps, and small vegetated patches, so our district-level summaries inevitably smooth over sub-pixel heterogeneity. Second, the single-channel algorithm for Landsat-derived land-surface temperature does not explicitly account for variations in atmospheric water vapour; while validation with AWS stations shows errors small enough for comparative district analysis, absolute temperature estimates may carry a non-negligible bias. Third, because Landsat over-passes occur in mid-afternoon, the diurnal cooling patterns that unfold later in the evening are not captured; future work (Sect. 5) should integrate higher-frequency thermal data (e.g. from geostationary sensors) to resolve these dynamics and better inform cooling-strategy design.

3.8 Downstream Use

The CSV output feeds Sect. 4, where districts with LST_C_p50 > 42 °C and NDVI_p50 < 0.15 are flagged for priority biophilic interventions. Cross-plots of NDVI versus NDBI reveal four district typologies—green-dense, mixed, grey-dense, and high-built—each matched with tailored, water-sensitive streetscape strategies.

4 Discussion and Implications

4.1 Why NDVI?—Theory, Math, and Measurement

Figure 2 illustrates a typical vegetation reflectance curve, highlighting the red (~650–680 nm) and near-infrared (~800–880 nm) bands used in NDVI computations. The Normalized Difference Vegetation Index (NDVI) converts raw satellite

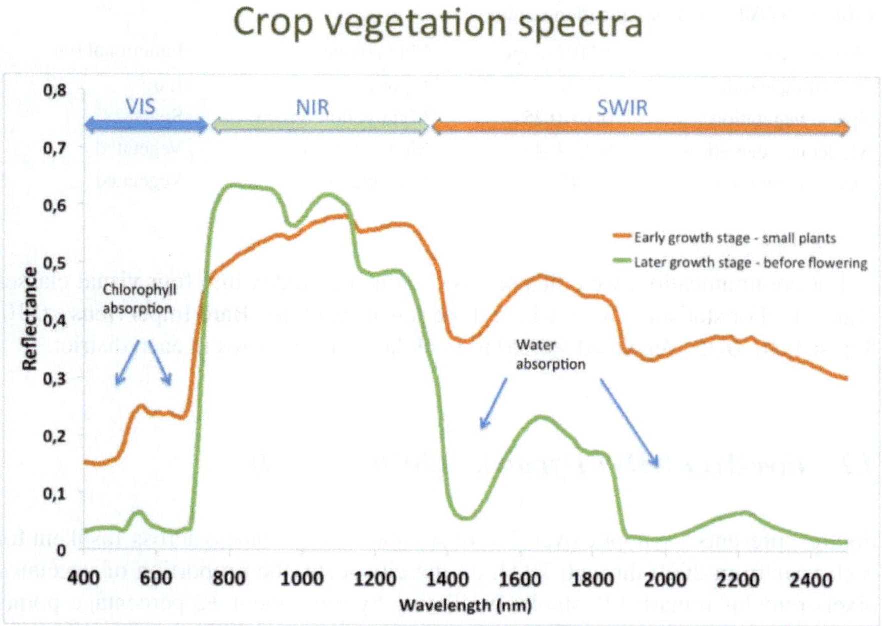

Fig. 2 Illustration of a typical vegetation reflectance curve (adapted from NASA/USGS spectral library)

reflectance into a unit-less proxy for "greenness". It exploits the strong contrast between chlorophyll absorption in the red band (R) and leaf-cell internal scattering in the near-infra-red band (NIR):

$$NDVI = \frac{(\rho_{NIR} - \rho_R)}{(\rho_{NIR} + \rho_R)}$$

where ρ is surface reflectance (0–1).

- *Physical meaning*—Healthy foliage (high NIR, low R) yields NDVI ≈0.6–0.9; bare soil clusters near 0 ± 0.1; water and impervious surfaces trend negative.
- *Sensor choice*—Sentinel-2 MSI (10 m) and Landsat-8/9 OLI (30 m) offer red/NIR bands that are radiometrically stable and free for global research—ideal for multi-year monitoring at city scale.
- *Assumptions and caveats*—NDVI saturates in dense canopies (>0.8), is sensitive to soil brightness in sparsely vegetated scenes, and must be atmospherically corrected. Using summer "peak-greenness" composites minimizes phenological noise in a dry-continental climate where vegetation is dormant for much of the year.

Table 1 NDVI visual classification scheme

Visual class	NDVI range	Map colour	Functional bin
Bare/Impervious	<0.10	Light grey	Bare
Sparse vegetation	0.10–0.25	Light yellow-green	Sparse
Moderate vegetation	0.25–0.45	Medium green	Vegetated
Dense vegetation	≥0.45	Dark green	Vegetated

For communication, we collapsed the continuous index into four visual classes (Table 1). For statistics, we retained three functional bins (Bare/Impervious <0.10, Sparse 0.10–0.25, Vegetated ≥0.25) to track land-cover shares in each district.

4.2 Five-Year NDVI Dynamics (2020 → 2024)

Figure 3 presents a synoptic overview of summertime greenness across Tashkent for each year from 2020 through 2024. At the city scale, the proportion of vegetated pixels remains remarkably stable, oscillating by only about ±2 percentage points around the long-term mean. This modest inter-annual variability appears linked to subtle shifts in precipitation patterns and irrigation scheduling rather than any sustained upward or downward trend in overall canopy cover.

Beneath this general stability, however, spatial heterogeneity is evident. The south-eastern agricultural fringe—comprising irrigated orchards and field plots—consistently shows the highest share of dense vegetation. In years when harvest or pruning occurs earlier, many pixels momentarily drop from the "dense" to "moderate" class, producing the darker patches observable in the 2021 and 2022 composites. These peripheral fluctuations account for the bulk of year-to-year NDVI noise.

Within the urban core, districts such as Mirzo Ulugbek, Yakkasaroy, and Shaykhontokhur maintain a stable, dense street-tree canopy, while private gardens and turf areas continue to dominate the "sparse vegetation" class. Conversely, modest infill development along major arterials has driven a small but noticeable expansion of bare or impervious surfaces—visible as light-grey swaths in the 2023 and 2024 maps—underscoring the tension between greening and densification pressures in the city centre.

The newly formed Yangihayot district (established in 2021) illustrates both sides of urban change. Initially characterized by a high bare-land share—reflecting its industrial estates and construction sites—it nevertheless registers the fastest absolute gain in dense vegetation by 2024, following the installation of shelterbelts and linear plantings along newly paved avenues.

Finally, districts with long-standing municipal greening programmes, notably Yunusobod and Olmazor, show gradual but consistent conversion of sparse-vegetation pixels into the vegetated class between 2022 and 2024. This "begrudging greening" appears as a fine, dark-green netting of moderate-vegetation over

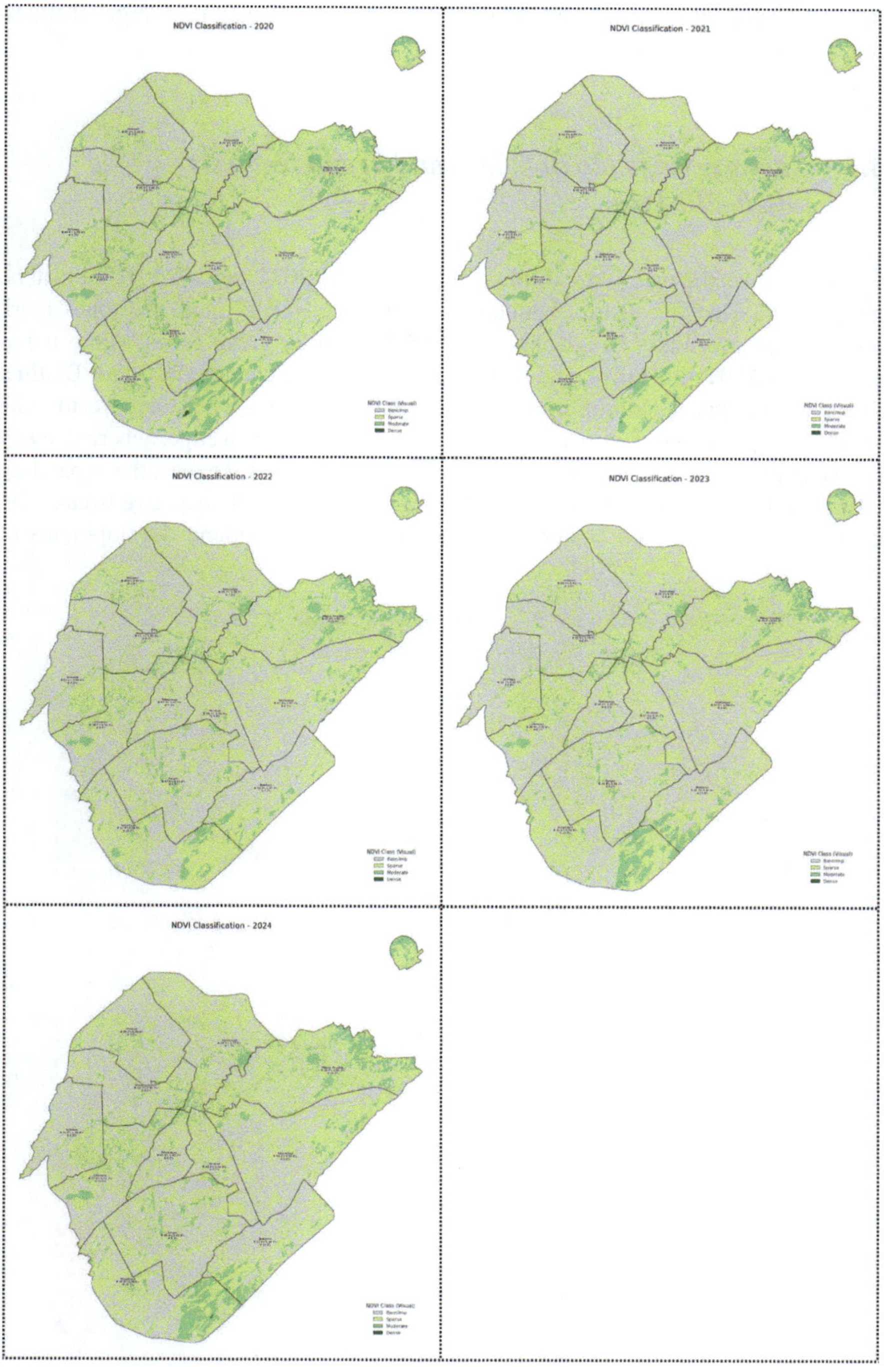

Fig. 3 Maps for summertime greenness in Tashkent for the period 2020–2024

previously uniform sparse areas, signalling the cumulative payoff of targeted street-tree and pocket-park initiatives.

4.3 *Change-Maps (2020–2024) and Trend Analysis (2020–2024)*

Figure 4 shows the pixel-wise difference between the 2024 and 2020 NDVI rasters, clipped to Tashkent's district boundaries. In this "ΔNDVI" map, red tones mark areas where greenness has declined (ΔNDVI < −0.1), green tones where it has increased (ΔNDVI > +0.1), and near-white where little change occurred. Distinct loss clusters appear along Sergeli's north–south arterials—where new residential infill has replaced street trees—and in the southern airport wedge, where runway resurfacing has eliminated turf and low-growing plantings. Likewise, the expanding Qibray limestone quarry at the city's edge is evident as sharply negative streaks. By contrast, prominent gain hotspots include the irrigated parkland developments of

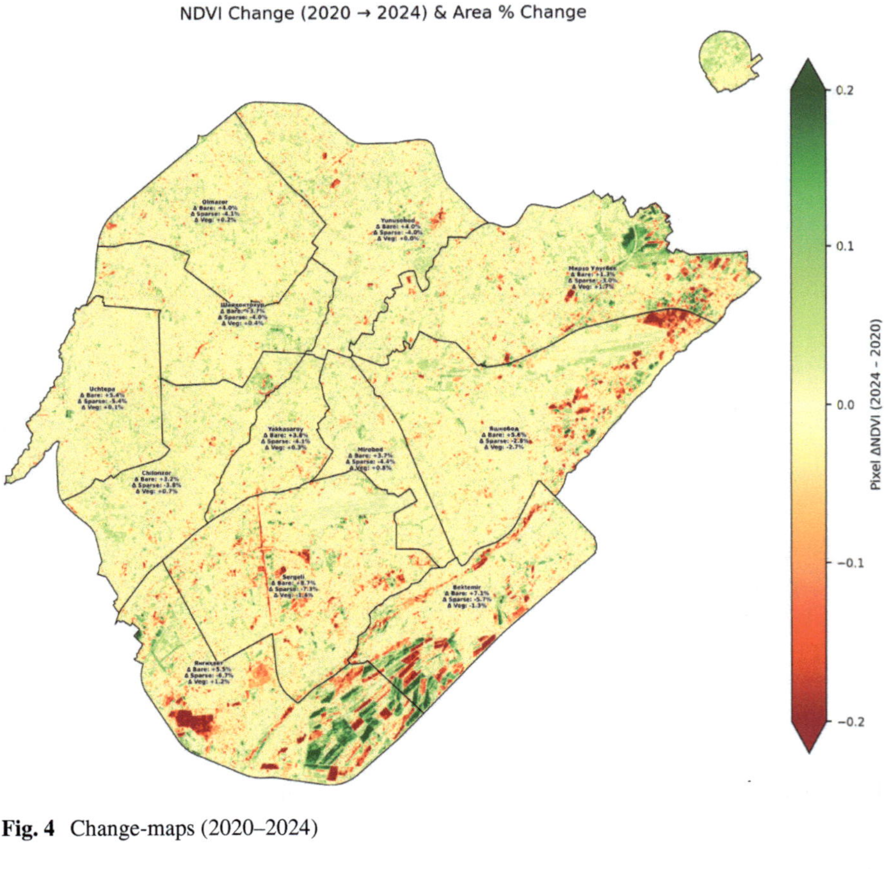

Fig. 4 Change-maps (2020–2024)

Mirzo Ulugbek in the northeast, shelterbelt plantings edging Bektemir's peri-urban agglomeration, and emergent wetland restoration projects just southeast of Yangihayot.

Beyond these raw change patterns, we quantified per-district trends in three functional land-cover categories (Bare/Impervious, Sparse Vegetation, and Vegetated) using linear regression on the annual percentages from 2020 through 2024. Most central districts (e.g. Mirzo Ulugbek, Yakkasaroy) exhibit a small but steady rise in the bare/impervious share (+0.5% per year on average) coupled with a corresponding decline in sparse vegetation (−0.4% per year). Vegetated cover trends are more variable: districts with active greening programmes (Olmazor, Yunusobod) show modest positive slopes (+0.2% per year), while rapidly densifying sectors (Sergeli, Bektemir) record slight negative or flat trends. These trend lines, plotted in Fig. 5, reinforce the ΔNDVI findings by demonstrating which districts are losing green "fabric" most rapidly and which are successfully converting sparse planting into sustained canopy cover over time.

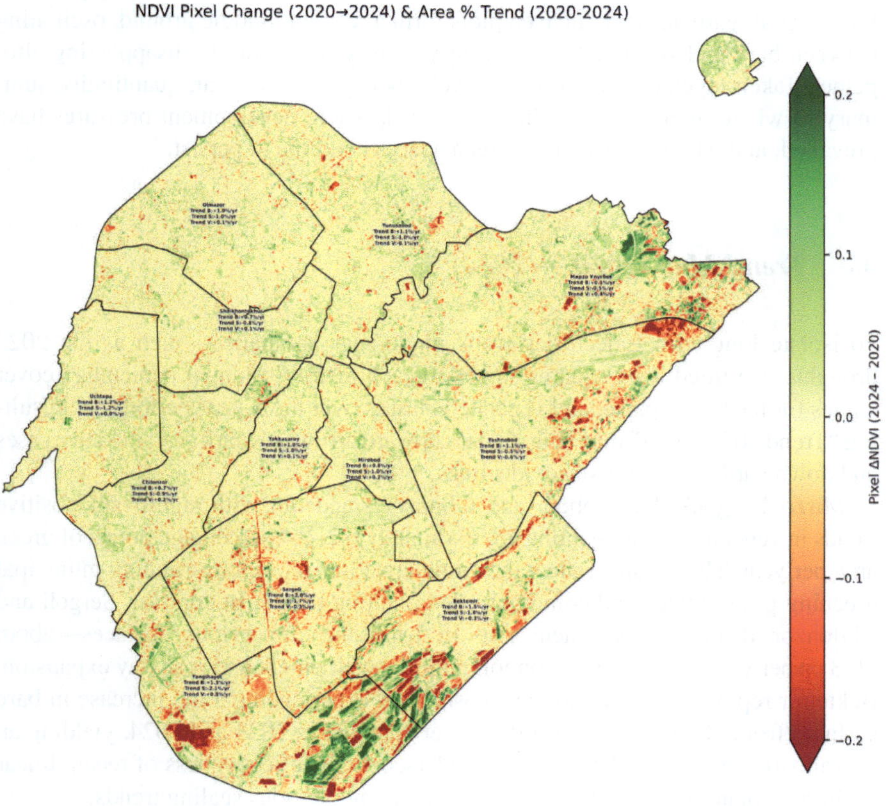

Fig. 5 Trend-map (2020→2024)

4.3.1 Area-Percentage Change Labels

In addition to the pixel-level ΔNDVI map, we calculated the change in areal share of the three functional classes—Bare/Impervious, Sparse Vegetation, and Vegetated—for each district between 2020 and 2024. Two thirds of the city's districts (8 out of 12) registered a net gain in vegetated cover, ranging from +0.3 to +1.4 percentage points. Bektemir led this city-wide greening trend, adding 1.4 percentage points to its vegetated share while simultaneously shedding 2.7 points of bare surface.

Conversely, the densest urban districts—Chilonzor and Uchtepa—experienced the opposite pattern: each lost nearly 0.9 percentage points of vegetated cover, a decline that was matched almost point-for-point by increases in bare and impervious surfaces. These shifts corroborate the red loss clusters in the pixel-difference map, underscoring where infill development has most aggressively replaced green fabrics.

Across all districts, the sparse-vegetation class remained remarkably stable, with absolute changes under 0.3 percentage points. This relative constancy suggests that lawns, small gardens, and unpaved plots form a resilient middle ground, oscillating between bare and vegetated states from year to year but rarely disappearing altogether. Taken together, these district-level labels provide a clear, quantitative summary of where urban greening has succeeded, where development pressures have prevailed, and where intermediate green spaces continue to persist.

4.4 Trend-Map (2020 → 2024)

To isolate long-term trajectories from single-year anomalies—such as the 2022 drought—we fitted least-squares slopes to each district's annual percentage cover values for the bare, sparse, and vegetated classes over the 5-year period. The resulting "Trend 2020–2024" map (Fig. 5) uses upward arrows to mark sustained increases and downward arrows for steady declines.

Mirzo Ulugbek, Yunusobod, and Olmazor stand out with significant positive trends in vegetated cover, gaining approximately 0.2–0.4 percentage points of green area per year. These gains reflect the cumulative impact of long-running municipal greening programmes and community-led planting drives. In contrast, Sergeli and Chilonzor display a persistent rise in bare and impervious surfaces—about +0.3 pp per year—indicative of ongoing infill development and roadway expansion. Bektemir represents a "turnaround" case: after experiencing a net increase in bare surfaces from 2020 to 2022, it shifts to net greening in 2023 and 2024, yielding an overall slope near zero. This reversal underscores the effectiveness of recent linear shelterbelt plantings in halting, and then reversing previous sealing trends.

By filtering out year-to-year noise and focusing on multi-year slopes, the trend-map provides a clear diagnostic of which districts are steadily greening, which are persistently sealing, and where targeted interventions have yielded measurable returns. This information is invaluable for allocating resources and setting realistic objectives for future biophilic streetscape projects.

4.5 Implications for Policy and Urban Planning in Dry-Continental Climates

In water-scarce, arid-continental settings such as Tashkent, a targeted micro-forestry strategy can deliver the most "greening per drop". By using our NDVI hotspot maps to identify parcels where dense-vegetation gain is greatest, planners can concentrate tree-planting and irrigation on those locations, thus maximizing cooling and biodiversity benefits while minimizing overall water demand.

Beyond tree planting, our pixel-difference analysis supports impervious-surface auditing. Red pixels on the ΔNDVI map directly flag new or expanded construction footprints. When these change detections are overlaid with cadastral permit records, municipal authorities gain a rapid-response tool to verify compliance with green-space requirements and to enforce mandated offsets when developments encroach on critical green corridors.

At the district scale, we advocate establishing interactive "greening dashboards" that report key performance indicators—such as annual percentage-point change in vegetated cover—side by side for all 12 districts. These dashboards enable decision-makers to allocate maintenance budgets and tree-planting resources equitably, directing additional support to districts where vegetation trends remain flat (under +0.1 pp per year) or are declining.

The co-benefits of climate adaptation are equally compelling. Literature from similar semi-arid cities indicates that each percentage point increase in vegetated cover can reduce peak surface temperatures by 1–3 °C. By aligning new plantings with high-density pedestrian corridors—especially the biophilic walking routes along Tashkent's canals—urban planners can significantly enhance thermal comfort, reduce heat-related health risks, and strengthen resilience to increasingly frequent heatwaves.

Finally, for monitoring frequency, we recommend a biennial cycle for full NDVI compositing to capture planning-scale trends while avoiding the cloud and atmospheric noise that can confound annual analyses. Nevertheless, annual pixel differencing remains a valuable rapid-response diagnostic for detecting illegal clear-cuts or construction activity, enabling authorities to intervene swiftly when green-space losses threaten the city's biophilic aspirations.

4.6 *Best-Practice Recommendations*

To ensure robust, context-appropriate green-cover mapping in arid-continental cities, we recommend adopting four complementary practices. First, always construct your annual composite from the same 6-week window each summer; this seasonal consistency removes phenological bias and isolates genuine land-cover change. Second, apply a morphological opening at the 30 m pixel scale before classification to suppress "salt-and-pepper" noise—isolated misclassified pixels that exaggerate bare-land estimates in coarser rasters. Third, calibrate your "dense vegetation" threshold to an NDVI of 0.45 or higher; in water-limited regimes, lower cutoffs tend to overestimate canopy cover and mask areas under moisture stress. Fourth, combine NDVI with the Normalized Difference Built-Up Index (NDBI) to distinguish bright impervious surfaces (e.g. concrete roofs) from dry bare soil, thereby improving the accuracy of bare-versus-built-up area statistics.

Beyond technical refinements, we advocate publishing interactive district-level dashboards that report both current vegetation percentage and multi-year trends. By making these metrics publicly accessible, municipalities can promote healthy inter-district competition, enable community groups to hold planners accountable, and streamline the allocation of greening resources where they are most needed.

4.6.1 Case-Study Reflection

Tashkent's modest net gain of +0.9 percentage points in vegetated area over 5 years illustrates that, even under severe water constraints, targeted urban-forest policies can counterbalance densification pressures. Scaling this GIS-driven workflow across other Central Asian cities would create a regional evidence base for dry-climate greening strategies and support coordinated, place-based interventions that enhance resilience and well-being.

5 Concluding Remarks

5.1 *Synthesis of Main Findings*

Our multi-year NDVI assessment reveals several interlocking insights about Tashkent's urban greening dynamics. First, the city's pattern of vegetative cover is closely linked to its historic canal network and associated green patches, with consistently higher NDVI values detected within corridor buffers than in adjacent urban fabric. Second, although the overall proportion of vegetated land within each district has changed only modestly between 2020 and 2024, the spatial distribution of gains and losses pinpoints precise neighbourhoods—such as peri-urban agricultural fringes and newly infilled central streets—where targeted greening interventions can yield the greatest benefit. Finally, when we overlaid the officially designated biophilic walking routes onto our 2024 NDVI map, a striking alignment emerged: almost every route follows the canals, underscoring these waterways' continuing role as Tashkent's primary green backbone and suggesting that future corridor-scale planning should build on this existing spatial logic.

5.2 *Limitations*

Several caveats temper our conclusions and point the way to future research. Our reliance on NDVI as a single greenness metric, while expedient, overlooks complementary signals that indices such as NDBI or EVI provide; integrating those and other environmental datasets (soil moisture, surface temperature) would offer a more holistic picture of urban ecosystem health. The temporal resolution of one summer composite per year also misses off-peak phenological phases—spring blooms and autumn senescence—that influence human well-being and thermal comfort. Spatially, summarizing at the district level smooths over fine-scale heterogeneity in private gardens, pocket parks, and street trees, limiting our ability to design hyper-local interventions. In addition, gaps in ancillary data—especially uniform NDBI coverage and high-resolution land-use maps—constrained our socio-ecological modelling. Finally, the set of biophilic walking routes used in this study derives from formal recreational maps and may under-represent informal pedestrian pathways; future work should incorporate participatory GPS tracking or crowd-sourced route mapping to capture the full spectrum of everyday green-space use.

5.3 *Brief Description of the Biophilic Routes Map*

The accompanying Fig. 6 presents the network of Tashkent Biophilic Walking Routes (May 2025) superimposed on the city's canal system and the 2024 NDVI distribution. Green-shaded corridors highlight the highest-greenness zones, while bold black lines delineate district boundaries. Importantly, local stakeholders—including Mahalla committee representatives, urban park user groups, and municipal planners—were convened in a participatory GIS workshop (March 2025) to validate and refine these routes, ensuring that the final alignment reflects both empirical NDVI "hotspots" and community-identified priorities (see Table 2).

Below is a proposed Table 2, showing how each stakeholder group contributed to the selection and validation of the biophilic walking routes, followed by a brief description of how these roles and outputs support our methodology and results.

In crafting our biophilic route network, we deliberately wove together quantitative GIS analytics and rich local knowledge to validate and refine each step of the process. By overlaying the Research Team's NDVI hotspot maps with hand-drawn

Fig. 6 Tashkent Biophilic Walking Routes (May 2025)

Table 2 Stakeholder roles and outputs in biophilic route selection

Stakeholder	Role/activity	Method of engagement	Deliverable/output
Municipal Planning Department	Provided base GIS data and policy guidance	Data-sharing workshop	Updated street network and canal vector layers; policy brief
Local Communities (Mahalla Committees)	Ground-truthing routes; identifying preferred green spaces	Participatory mapping sessions	Hand-drawn route sketches; qualitative feedback report
University Research Team	Conducted multi-temporal NDVI and ΔNDVI analyses	Joint field excursions; online dashboards	District-level NDVI statistics; pixel-map layers (GeoTIFFs)
Landscape Architects	Translated GIS outputs into schematic designs	Design charrette	Concept sketches for canal-buffer planting; species palettes
Environmental NGO	Facilitated community workshops; raised awareness	Public outreach events	Workshop summary with >50 resident interviews; photo documentation
Citizen-Science Volunteers	Collected on-site microclimate measurements	Mobile app data collection	200+ temperature and soil-moisture points; CSV time-series

route sketches produced by Mahalla Committee members, we confirmed that the high-greenness corridors detected via remote sensing correspond directly to the parks and tree-lined paths that residents frequent. In one striking instance, the dense-vegetation cluster highlighted in Mirzo Ulugbek's NDVI composite coincided exactly with a canal-side pocket park that community participants had flagged as a favourite gathering place.

At the same time, the Municipal Planning Department translated these spatial findings into policy guidance. Their brief synthesized metrics—such as the districts showing more than a one-percentage-point gain in vegetated cover—into clear targets for upcoming tree-planting budgets. They also updated the city's GIS layers so that any future development permit automatically flags canal buffer zones for priority greening, ensuring our analysis feeds directly into planning workflows. The collaboration continued through design: landscape architects convened in a charrette to convert our raster outputs into concrete planting schemes. One proposal—alternating native shade trees every 20 m along primary canal trails—emerged from these sessions and was then presented in a second community workshop. Feedback from residents refined the spacing, species mix, and path alignments, tightening the fit between "big-data" insights and cultural preferences on the ground.

Meanwhile, citizen-science volunteers collected more than 200 microclimate measurements on site, confirming that shaded canal sections run about 2 °C cooler than unshaded stretches. These empirical data points not only reinforced the pixel-level ΔNDVI loss and gain patterns observed from 2020 to 2024 but were also fed back into the GIS model to fine-tune the final corridor selection. Together, these intertwined contributions—spatial analytics, policy translation, design iteration, and community-driven validation—yielded a robust, co-created biophilic route network. The resulting map is grounded in hard data yet alive with the lived experiences and aspirations of Tashkent's residents.

Figure 7 lays out the six-stage process through which raw remote-sensing data are transformed into a fully co-created, policy-ready biophilic walking-route network for Tashkent. The workflow begins with NDVI corridor detection, in which summer composites of Landsat and Sentinel imagery are classified and thresholded to reveal the city's greenest linear features—those stretches of high normalized difference vegetation index that promise the greatest biophilic benefit.

These candidate corridors then enter a stakeholder validation phase, where local residents, mahalla committees, and municipal planners review the remotely-sensed "hot-spots" against lived experience and local knowledge. Feedback from community workshops and on-site field visits helps to confirm which canal-adjacent parklands, street-tree boulevards, or neighbourhood greenspaces truly matter to users.

Validated corridors are next integrated back into the municipal GIS layers in a policy integration & GIS update step. Here, greening targets derived from the NDVI statistics—such as districts exhibiting more than a 1%-point rise in vegetated cover—are converted into draft amendments to Tashkent's tree-planting budgets and zoning regulations. The updated GIS database ensures that all future permit reviews will automatically flag these priority corridors for preservation or enhancement.

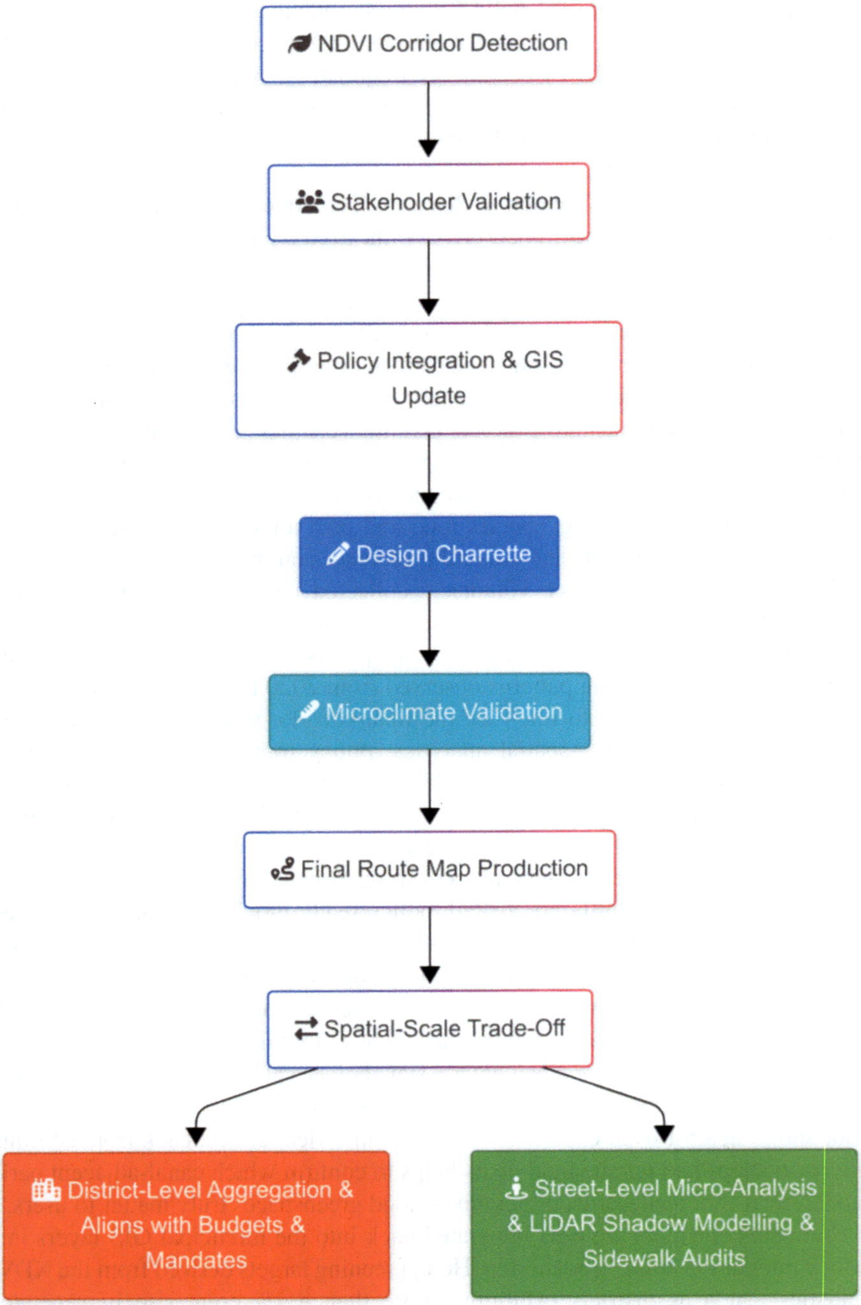

Fig. 7 Process flow for co-creating and validating Tashkent's biophilic walking-route network

With policy traction secured, a design charrette convenes landscape architects, urban designers, and community representatives to translate GIS outputs into concrete streetscape proposals: specifying tree species, planting intervals, path alignments, and site furniture. These schematic designs then undergo microclimate validation, in which portable sensors and volunteer measurements confirm that recommended shade-tree patterns indeed reduce air and surface temperatures by 1–3 °C along critical canal segments.

Finally, the refined proposals are compiled in the final route map production, a cartographic deliverable that overlays the biophilic network onto Tashkent's administrative districts, complete with annotated photograph insets and a full legend.

At the end of the flow, a spatial-scale trade-off node highlights two complementary paths forward. On one hand, district-level aggregation aligns the network with existing budgeting structures and greening mandates, enabling straightforward appropriation of staff time and funding. On the other, street-level micro-analysis—drawing on LiDAR shadow models and sidewalk-width audits—offers a higher-resolution toolkit for targeting planting sites along narrow alleys, parklets, and individual blocks. Together, these branches ensure that the biophilic framework remains both administratively feasible and sensitive to fine-grained local contexts.

6 Practical Pathways for Scaling the Biophilic Framework

To extend the lessons learned in Tashkent to other major cities across Central Asia—such as Samarkand, Bukhara, and Namangan—it is essential first to establish a formal network for collaboration. By convening a Central Asian Green Cities Consortium, municipal authorities and urban planners can share proven GIS-based workflows, NDVI/NDBI baselining methods, and model policy templates for designating "biophilic corridors". Such a consortium would serve as a living repository of best practices, enabling peer-to-peer learning and harmonized standards for corridor identification, mapping, and management.

Equally important is the integration of biophilic metrics into each city's master planning framework. This requires amending urban development regulations to embed quantitative greening targets—for example, mandating a minimum green-to-grey surface ratio (such as 1:3 vegetation to impervious cover) along all primary arterials. By leveraging satellite-derived indices to measure compliance, and by presenting these metrics on transparent, district-level dashboards, planners and elected officials can track progress, allocate maintenance resources strategically, and tie tree-planting budgets directly to measurable outcomes.

Finally, truly resilient and community-driven biophilic networks depend on the active participation of residents. Deploying simple mobile or web-based applications (such as Maptionnaire) allows local stakeholders to report heat-stress locations, flag preferred walking routes, and submit real-time microclimate observations. These crowd-sourced data streams can be ingested into ongoing NDVI monitoring and route-planning workshops, creating a feedback loop that continually refines both the spatial configuration of green corridors and the selection of species or design interventions best suited to on-the-ground conditions.

By weaving together remote sensing analytics, participatory mapping, and policy integration, this three-pronged roadmap offers a clear, replicable model for arid-continent cities seeking to embed nature into the urban fabric. In doing so, it advances not only thermal comfort and ecological connectivity but also the cultural vitality and well-being of communities across the region.

The collective evidence from our mapping and community validation exercises underscores the strategic value of Tashkent's canal corridors as the city's green backbone. Decision-makers are therefore urged to focus their greening efforts on these riparian zones by selecting native, drought-tolerant shade trees that thrive in the local climate and by encouraging a richly layered understory of shrubs and grasses. Establishing uninterrupted belts of vegetation along both banks will not only reinforce ecological connectivity and urban cooling—mitigating the summer heat island effect—but also enhance biodiversity by creating habitat linkages for birds, pollinators, and small mammals. Well-designed canal buffers can double as high-quality public realms, offering shaded walkways, seating niches, and seasonal blooms that foster daily restorative experiences for residents. By embedding these continuous green ribbons into municipal planning, Tashkent can build resilience to rising temperatures and water stress while delivering long-term health, social and cultural benefits to its communities.

References

Appleton, J. (1975). *The experience of landscape*. Wiley.

Dramstad, W. E., Olson, J. D., & Forman, R. T. T. (1996). *Landscape ecology principles in landscape architecture and land-use planning*. Island Press.

Forman, R. T. T., & Godron, M. (1986). *Landscape ecology*. Wiley.

Gómez-Baggethun, E., Gren, Å., Barton, D. N., Langemeyer, J., McPhearson, T., O'Farrell, P., et al. (2013). Urban ecosystem services. In T. Elmqvist, M. Fragkias, J. Goodness, B. Güneralp, P. J. Marcotullio, R. I. McDonald, et al. (Eds.), *Urbanization, biodiversity and ecosystem services: Challenges and opportunities* (pp. 175–251). Springer. https://doi.org/10.1007/978-94-007-7088-1_11

Intergovernmental Panel on Climate Change. (2021). *Climate Change 2021: The physical science basis. Contribution of Working Group I to the Sixth Assessment Report of the Intergovernmental Panel on Climate Change* (V. Masson-Delmotte, P. Zhai, A. Pirani, S. L. Connors, C. Péan, S. Berger, ... B. Zhou, Eds.). Cambridge University Press. https://doi.org/10.1017/9781009157896

Kaplan, R., & Kaplan, S. (1989). *The experience of nature: A psychological perspective*. Cambridge University Press.

Kellert, S. R., Heerwagen, J. H., & Mador, M. (Eds.). (2008). *Biophilic design: The theory, science, and practice of bringing buildings to life*. Wiley.

Kellert, S. R., & Wilson, E. O. (Eds.). (1993). *The biophilia hypothesis*. Island Press.

Ryan, C. O., Browning, W. D., Clancy, J. O., Andrews, S. L., & Kallianpurkar, N. B. (2014). Biophilic design patterns: Emerging nature-based parameters for health and well-being in the built environment. *ArchNet-IJAR: International Journal of Architectural Research, 8*(2), 62.

Sharipov, S., & Khayitmurodov, A. (2024). *The impacts of green spaces on mitigating the urban hot island effect in the city of Tashkent.* BIO Web Conf. 105 06013. https://doi.org/10.1051/bioconf/202410506013

Terrapin Bright Green. (2014). *14 patterns of biophilic design: Improving health & well-being in the built environment.* Terrapin Bright Green, LLC. https://www.terrapinbrightgreen.com/reports/14-patterns/

Tzoulas, K., Korpela, K., Venn, S., Yli-Pelkonen, V., Kaźmierczak, A., Niemelä, J., & James, P. (2007). Promoting ecosystem and human health in urban areas using green infrastructure: A literature review. *Landscape and Urban Planning, 81*(3), 167–178. https://doi.org/10.1016/j.landurbplan.2007.02.001

Ulrich, R. S. (1984). View through a window may influence recovery from surgery. *Science, 224*(4647), 420–421. https://doi.org/10.1126/science.6143402

Ulrich, R., Simons, R., Losito, B., Fiorito, E., Miles, M., Zelson, M. (1991). Stress recovery during exposure to natural and urban environments. *Journal of Environmental Psychology, 11*, 201–230. https://doi.org/10.1016/S0272-4944(05)80184-7

UNDP Report. (2023). *The future of green transition in Uzbekistan.* https://www.undp.org/sites/g/files/zskgke326/files/2023-11/future_of_green_transition_eng.pdf#page=4.15

Wilson, E. O. (1984). *Biophilia.* Harvard University Press.

Heritage, Identity and Urban Transformation

Genova Behind the Scenes: A Didactic Approach for Including Urban Nature in Placemaking-Heritage Driven Strategies

Israa Mahmoud, Zachary M. Jones, Abdallah Jreij, and Necdet Ayik

Abstract With the increased understanding of climate-change-related pressures facing cities today, the need to include nature-based solutions (NBS) in urban areas has become urgent. Aligning with the EU Biodiversity Strategy 2030 (hereafter EUBDS2030), several 20K-inhabitant cities have been striving to design and adopt urban greening or urban nature plans as part of their climate adaptation strategies. These approaches often include biophilic design and NBS as part of spatial strategies, which can introduce many benefits of nature in dense built environments. However, such plans also bring challenges in terms of the economic, social, and cultural aspects of cities. These challenges may be better addressed when participatory design mechanisms are introduced into the process to alleviate municipal responsibilities.

This research paper focuses on exploring potential methods to support the adoption of the Genova Green Strategy, launched in 2022, as a guiding document for enhancing the green and blue infrastructure of the city. Based on this goal, we unpack the possibility of using NBS in improving urban design strategies at the local level while activating placemaking strategies to enhance community engagement with heritage embedded in Genova's historic fabric.

Through an experimental didactic approach that intersects three main topics of placemaking, NBS and urban heritage, this study reflects on the outcomes of an urban design studio involving 50 students focused on the city of Genova in Italy. Through a three-phase process, students were introduced to the concepts of NBS and biodiversity planning within urban heritage contexts, then to placemaking approaches and their different implementation techniques. Lastly, students used stakeholder mapping and possible urban design scenarios to develop a better understanding of nature-driven planning in UNESCO World Heritage sites. This

I. Mahmoud (✉)
Laboratorio di Simulazione Urbana (Fausto Curti), Department of Architecture and Urban Studies, Politecnico di Milano, Milan, Italy
e-mail: israa.mahmoud@polimi.it

Z. M. Jones · A. Jreij · N. Ayik
Department of Architecture and Urban Studies, Politecnico di Milano, Milan, Italy

A. Cheshmehzangi et al. (eds.), *Designing Nature into Cities*, Urban Sustainability, https://doi.org/10.1007/978-981-95-9382-8_5

experimental approach addresses a key gap in current research by exploring the intersection of these three areas and by demonstrating the potential for implementing NBS through placemaking strategies in heritage-rich urban contexts.

Keywords Placemaking · Nature-based solutions · Urban design · Didactic approach · Heritage · Green strategy

1 Introduction

Research on nature-based solutions (NBS) has advanced significantly over the past decade, following the launch of the standardized definition by the International Union of Conservation of Nature (IUCN, 2020). This was later further developed by the UN Environmental Assembly resolution 5/5 as actions to protect, conserve, restore, and sustainably manage natural or modified ecosystems (Convention on Biological Diversity, 2021; UNEP, 2022). NBS have proven to be an effective measure of adaptation and mitigation to climate change while working across different spatial scales such as strategic urban planning, whereas the diffusion of green corridors and larger-scale measures has demonstrated the benefits of increasing biodiversity (Mahmoud, 2022, 2024). Meanwhile, at the micro and meso spatial scales, challenges remain in identifying the right NBS for the specific type of challenge, such as green roofs and walls to improve the microclimate conditions at the building and the street scale (Kamjou et al., 2024; Kowarik et al., 2025).

However, the intersection of this topic in relation to other soft urbanism and planning topics such as placemaking approaches or issues of heritage conservation, especially in cities with diversified urban morphology where standardized solutions may not be appropriate, has been quite noticeable. The biophilic design approach, for instance, and how people interact with nature in public spaces have become a potential policy tactic to explore possible characteristics in which the urban environment can readapt to impact human health and well-being (Andreucci et al., 2021). Nonetheless, this concept of biophilic design is not the only example; more recently, nature-driven urbanism (Roggema, 2020) and technological NBS (Galle et al., 2019) have surfaced since the COVID-19 emergency locked in several populations around the world with even less—or no—access to natural environment in their immediate surroundings.

NBS, however, do not only intersect with environmental challenges but also have significant social and cultural benefits in cities, especially in vulnerable communities where economic capacities and financial resources are limited (Bockarjova et al., 2022). It is also evident that larger-scale cities or ones with particular urban morphologies such as coastal cities or heritage cities have a more pronounced need to use NBS not only for tackling climate resilience but also for enhancing their social cohesion and local communities' structures in times of flash floods, heat waves, or other climate emergency. Hence, public participation and co-creation notions can be integrated into the urban planning processes to enhance citizen engagement (Bradley & Mahmoud, 2024; European Commission, 2023). One possible pathway is to use micro-urban design strategies such as placemaking and tactical urbanism strategies that offer *"lighter, quicker, cheaper"* solutions and allow citizens the opportunity to amplify the impact of NBS at the local level. When mainstreamed and upscaled across the neighbourhoods at the meso-scale and the city as a whole, this can generate momentum for natural environments to thrive in cities.

Participatory design approaches, nonetheless, also require a standardized framework to operate within, especially when the contribution is related to public spaces and larger scales beyond the neighbourhood and the building micro-scales. This aligns with the latest EU Biodiversity Strategy 2030 (Viti et al., 2024), which mandates that cities with more than 20k inhabitants are required to adopt a green or nature plan on several scales to manage and monitor their progress towards the larger Global Biodiversity Framework (Bulkeley, 2020; CBD, 2022). Specifically, target 14 in the GBF calls for an inclusive urban planning in which citizens are made part of the governance processes for enhancing nature and biodiversity in their cities (Mahmoud et al., 2025).

This research paper attempts to bridge the gap between the theoretical knowledge of NBS and the practice of placemaking approaches from a didactic perspective using the city of Genova, Italy, as a case study. It adopts a three-phase approach to analyse, design, and assess the possibilities of enhancing urban assets using nature-driven placemaking strategies especially in a heritage context through which both urban morphology and heritage aspects were introduced and intrinsically challenged.

2 Background Knowledge About Genova as a Case Study

Genova is an important historic city in Italy, situated between the sea and the mountains. The city of Genova has a long history, pre-dating the Roman era, with its historical development and port largely informed by its geographical location. As one of the historic maritime republics operating in the Mediterranean, the city's most notable period occurred during the sixteenth and seventeenth centuries as an independent maritime republic. In 2006, part of the city centre was inscribed as a UNESCO World Heritage Site as Genoa: Le Strade Nuove and the system of the Palazzi dei Rolli. Through the 1990s and early 2000s the waterfront and historic city centre went through several phases of urban renewal and regeneration attached to the large-scale events of the Genova Expo '92, 2001 G8 Summit, and the European Capital of Culture 2004 (Jones, 2022). These works were critical to valorizing the city's rich history as well as making it more accessible to citizens and visitors alike. The city's heritage has played a key role in its recent urban development and can continue to play a key role in its future development.

Genova has also faced several severe environmental challenges in recent years, including flash flooding, landslides, and rising sea levels—in addition to a lack of green and public spaces to serve local communities. These events have been occurring at greater frequency over the last 10 years and several areas of the city, including a portion of the UNESCO World Heritage Site, are considered at risk of severe hydrological flooding events (Paliaga et al., 2018). The city's adjacent mountain terrain also puts the area at risk of landslides, with one of the nearby villages of 1000 people entirely cut off from the rest of the city in 2024. Natural disasters are occurring at an increasing rate, putting the city's fragile cultural heritage at ever greater risk. These issues present complex and challenging urban and environmental problems to be addressed. On one hand, there is an urgent need to ensure the protection of the city's cultural heritage while, on the other, there is the emerging need for new and innovative solutions to address these climate-change-driven crises affecting the urban landscape.

At an international level, there is growing concern and recognition of the need to address environmental threats facing cultural-heritage-rich cities such as Genova around the world. In 2019, ICOMOS recognized the need to overcome the nature-culture divide (ICOMOS, 2019), while the 2022 Climate Change Adaptation report highlights the many threats facing heritage cities from climate change (ICOMOS, 2022). The 2021 European Cultural Heritage Green Paper by Europa Nostra and ICOMOS also aims to highlight links between the ecological and social functions of heritage landscapes (Europa Nostra & ICOMOS, 2021). However, despite the efforts of these international documents to highlight the issues at hand, there remains a lack of specific targets to meet or recommend policy approaches to follow and implement regarding NBS or other approaches to re-introducing nature into historic environments.

Since the inception of the European Union Biodiversity Strategy 2030, several countries have worked closely on National Biodiversity Strategies to ensure a cohesive approach, setting plans and targets according to the mandate of the Nature Restoration Law, recently approved in June 2024. Cities have since been advancing urban regeneration plans and architectural projects that match these objectives of incrementing climate resilience and naturalization at the local level (SDG 13—Climate Action). The city of Genova, amongst others in Italy, has also joined the wave with Genova Green Strategy 2022 designed by *OpenFabric* that situates Genova at the centre of the Mediterranean and European landscape.

Indeed, the city "Action Plan Genova 2050 programmatic initiative" aims at resolving several challenges of territorial governance and cohesion through a shared and inclusive vision with innovative development of green, blue, and grey infrastructures, climate change, urban regeneration actions, as well as measures for social and economic resilience. More recently, in November 2024, the city of Genoa has managed to publish its first green plan[1] which incorporates the green strategy developed in 2022 with the rest of the executive planning of the municipality for the

[1] https://www2.comune.genova.it/content/piano-del-verde-0 (last Accessed November 2025)

Fig. 1 The six types of urban morphologies according to the classification of the city of Genova in 2022. *Source*: Used with certified/written permissions from OpenFabric https://www.openfabric.eu/projects/genova-green-strategy/

implementation of green areas in public spaces as a response to climate change and urban qualities as a new main regulatory framework for the city's different types of urban morphology, see Fig. 1.

For these reasons, Genova provides an ideal case for which to develop an experimental didactic approach to confront these challenges of climate change and cultural heritage preservation. Genova represents an important historic city that is facing extreme climate-change-related threats, and it stands out as one of the few historic European cities to have developed a green plan to help confront the many emerging problems. As there has been limited reflection on how to practically integrate green and NBS within historic areas, the context of Genova provides an ideal basis to explore these issues. Confronting both the protection of cultural heritage and the need to generate new urban nature solutions challenged students with a set of urgent real-world issues in which there are no pre-defined solutions to be readily implemented. Rather, the complex nature of the case required students to thoroughly analyse and understand the specificity of the context in which they were working to develop and propose tailor-made placemaking solutions.

Following an in-depth analysis of the case of Genova and future scenarios in which climate-change pressures continue to proliferate, we carry out a comprehensive investigation of how to effectively address nature-driven placemaking while accounting for local heritage. Finally, the research explores a collective framework for the nine municipalities that make up the city of Genova in 2050 through specific design proposals. This diffused approach provides a broad exploration of how these issues can be properly addressed, the challenges encountered, and the range of possible solutions which emerge.

3 Methodology

The Urban Design Studio is formed as an experimental studio exploring the intersection of placemaking strategies, NBS, and urban heritage. The process is constructed across multi-scalar phases between urban planning and urban design practices aligned with sustainable development goals. The workflow is structured in three stages, each concluding with public presentations by the student groups.

In the first task, each group of students was assigned a specific sub-municipality of Genova. The aim was to develop potential urban strategies and comprehensively analyse the urban components with particular attention to the selected contexts. This phase included an in-situ observational study, see Fig. 2. The site-study integrated both instructor-led sessions and independent exploration of the students. Furthermore, the process was enriched by the presentations of the guests who participated in the research and practices in similar urban contexts, specifically in Genova, alongside the teaching team's lectures that supported the process with theoretical and practical knowledge. A key moment in the presentation series was the organized seminar that hosted Francesco Garofalo, founder of *Openfabric* (an international landscape architecture and urbanism practice based in Rotterdam, Milan, and Genova). His presentation offered students a direct insight into the professional processes behind the development of Genova Green Strategy. Francesco explained that the strategy was conceived as an ambitious citywide initiative aimed at redefining the relationship between urban form and nature. He also detailed how they approached the city's complex morphological layers, environmental challenges, and socio-spatial dynamics to design a multi-scalar framework for urban greening, climate adaptation, and public space transformation.

Fig. 2 Collective site visit with students, professors, and instructors. *Source*: The authors

Due to the particularly complex geographical makeup of Genova, each group was required to investigate the unique range of issues which typify their sub-municipality. The coastal areas are confronted by rising sea levels while the more mountainous areas must deal with landslides. The flat portions of the city and those located near rivers and runoff areas are routinely flooded. While the historic city centre hosts the UNESCO-listed World Heritage Site, which clearly has a recognized universal value, students working in other areas of the city also investigated the wider tangible and intangible cultural heritage assets of the city. This provided the opportunity to go beyond the widely accepted definitions of heritage to embrace less visible or recognized heritage, from food traditions, villas, landscape heritage, and abandoned industrial sites. In this way, the analysis investigation was specific to each individual sub-municipality but collectively provided a complex and complete overview of the entire city.This phase was followed by the master planning proposals using insights gained during the site visit and preliminary reviews. Conceptual exploration and identification of the key challenges and needs guided the initial ideas, structured around student-selected thematic inputs. This period facilitated the integration of prior knowledge into the spatial context at the scale of the master plan.

Students were challenged to determine their own master plan locations based on the varied analyses carried out and through the use of decision-making tools such as SWOT, STEEP, and Stakeholder Mapping that helped them determine the most critical and opportune areas to be addressed. Rather than being provided with a predetermined area of intervention, students had to make strongly supported arguments for the selected areas to intervene, especially noting the presence of cultural heritage and opportunities to address environmental challenges. This added level of engagement required students to seriously confront the two main themes of the course and take responsibility for the development of their projects and how these would engage with local communities. Additionally, this method ensured that students actively referred to and utilized the analysis conducted in the first phase as it was critical for the selection of the areas to work in and the development of a strategic master plan.

The third section of the design studio was dedicated to "neighbourhood projects," in which students were tasked with developing site-specific urban interventions and projects. Particular emphasis was placed on incorporating placemaking principles, NBS, and urban heritage phenomena harmonized with the multi-scalar and time-based programmatic strategy. By considering the previous design decisions and studies, the students developed urban design proposals under regular supervision and revision sessions by the instructors.

The overall methodology forced students to work at multiple scales, from the entire metropolitan territory of the city down to individual neighbourhood-scale interventions that ranged from the design of new public spaces, gardens, and transport systems to community centres, cultural hubs, and sports complexes. The diversity of outcomes and design solutions is testament to the structured approach that forced them to work locally and respond to the individual needs and challenges of the diverse locations selected. A wide array of placemaking strategies were proposed that were integrated with biodiverse and NBS to provide new green spaces and amenities that reinforced local heritage values.

4 Discussion and Implications for Didactic Experimentation

Based on the division across the nine sub-municipalities of the city of Genova, the students developed nine individual master plans to address the specific needs of these areas, which varied from the historic centre to the coastal and rural mountain areas, see Fig. 3.

Across these nine master plans, 25 urban projects were elaborated and designed in detail. Each of these projects was tasked with implementing specific placemaking strategies that would introduce NBS as well as engage with the city's wide range of cultural heritage. Of these 25, we present 2 selected projects here to highlight the challenges faced by the students and how their analytical investigation and strategy development led them to devise specific and place-based design solutions.

Through their project for the Cornigliano neighbourhood, see Fig. 4, the students simultaneously addressed the issues of flood risk, abandoned areas, and the regeneration of the historic Villa Bombrini and Park. Their project reactivates these areas through the insertion of a new public space that adds new functions and increases the biodiversity of the area. The wetland park responds to the changing flooding conditions of the Polcevera Torrent. Their design unifies and provides new access to this area that was previously divided and inaccessible with limited uses. The design strategically introduces NBS into the existing cultural heritage of the Villa and Park with new wetlands, gardens, water-friendly plants, permeable pavements, and living

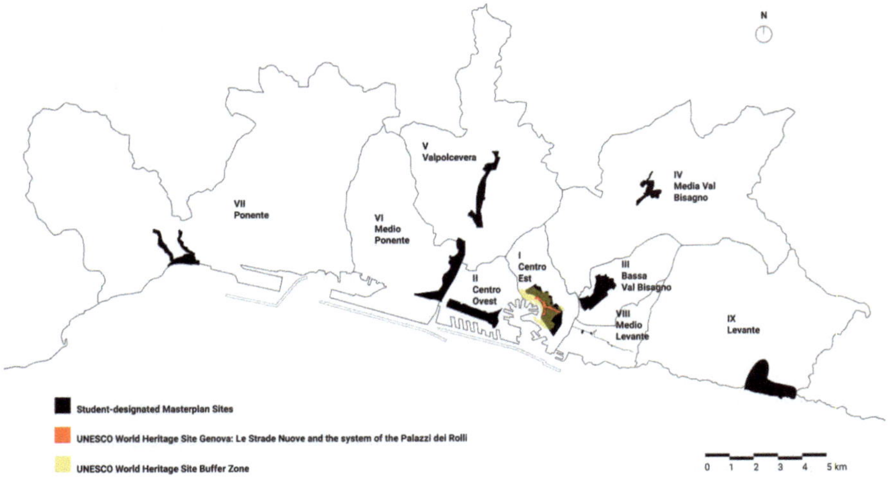

Fig. 3 The compilation of the overall nine sub-municipalities Masterplans spread across Genova. *Source*: The authors

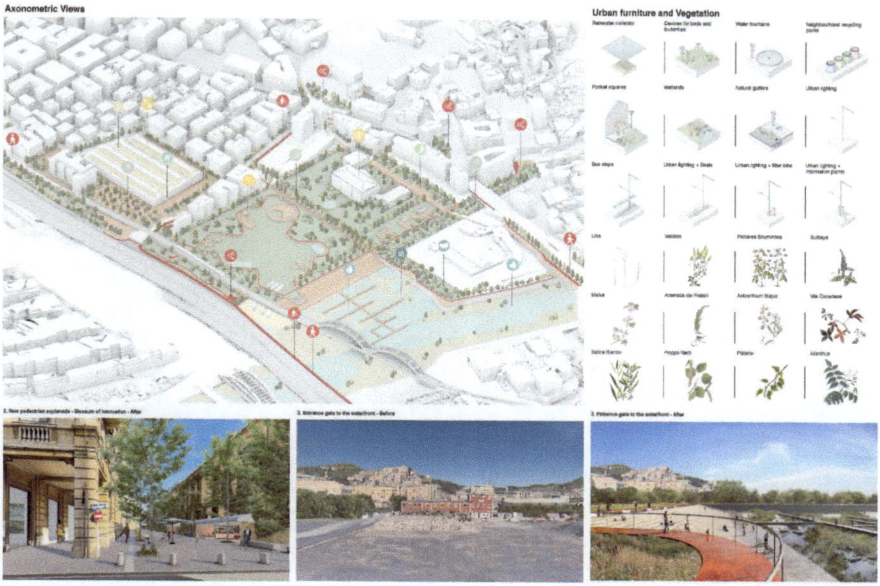

Fig. 4 Cornigliano neighbourhood project integrating flood mitigation, heritage regeneration and biodiversity enhancement through NBS around Villa Bombrini and its park. The developed axonometric view and the selected urban furniture and vegetation (top view) showcase the selection of place-based design solutions (bottom renders) as adequate for the nature-driven placemaking approach. *Source*: The authors

walls. These solutions are not in competition with the heritage site but rather will help to activate the area by making it an attractor of new uses for the local community and can also educate about the effects of climate change and the possible solutions.

Meanwhile in the mediaeval core of the city centre—see Fig. 5—which makes up part of the buffer zone of the UNESCO World Heritage Site, students implemented instead a series of micro solutions to respond to the lack of green spaces in the city centre. Many of the city's renowned Renaissance and Baroque palaces include urban green spaces and gardens, yet these have traditionally been exclusive spaces reserved for the privileged few. In this project, the students adopted and expanded this existing green heritage of the city through the insertion of new NBS throughout the historic centre. Based on careful analysis, they identified the urban spaces, streets, walls, and rooftops which could host a range of green solutions. Temporary and removable NBS included green shading, fountains, green roofs, planters, and self-supported green walls. These targeted urban micro solutions would address the issues of increasing temperatures and the urban heat island effect while also embracing the city's tradition of urban green spaces, interpreted through new design solutions. This project aims to achieve a nuanced balance between responding to emerging climate-change threats while also respecting and promoting existing heritage values.

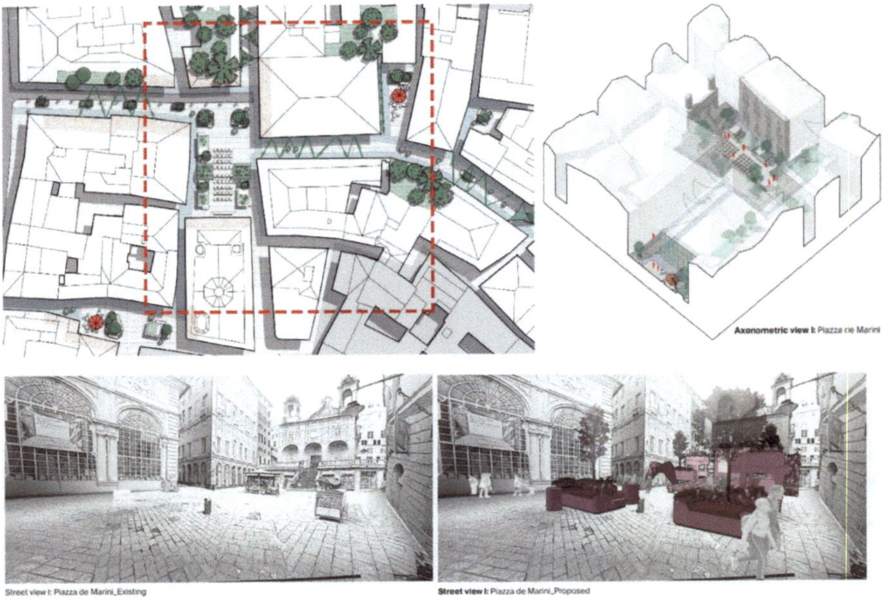

Fig. 5 The mediaeval city centre project proposing microscale NBS (e.g., green roofs, planters, and shading structures) to address the urban heat island effect. The Plan (upper-left) and Axonometric view (upper-right) explore possible integrations of these place-based solutions in the proposed piazza de Marini view (bottom-right). *Source*: The authors

5 Concluding Remarks

The didactic experiment provided students with a laboratory to investigate and respond to a complex set of urban issues that wider research and policy have not yet fully addressed. The examples from the students' works highlight first the possibility for NBS, placemaking, and urban heritage to intersect and support one another. The first example shows how these design approaches can be used to revitalize overlooked and undervalued heritage areas, incorporating them into long-term sustainable development goals. The second example highlights how microscale temporary interventions could come to be used to meet diverse needs, from activating historic spaces through new uses and providing much-needed green space. The wide-reaching approach that covered the entire metropolitan territory of the city also led to the development of a wide set of site-specific solutions that differed from area to area depending on the need of each site.

As an initial experiment that combines education activities with research goals, the work highlights the great potential of this approach as well as future aims to target. There are clearly a wide range of possible ways to engage and use NBS and placemaking strategies within heritage contexts. However, further research is necessary to fully address and understand these issues. The collection and analysis of existing examples will provide further evidence as to the existing state-of-the-art regarding what kinds of NBS projects have been proposed or implemented within heritage contexts. Further research is also needed to identify which range of solutions can be appropriate for different kinds of heritage settings, as not all possible interventions are necessarily adaptable to each kind of heritage context. Further reflection, study, as well as the development of initial pilot projects can begin to establish the necessary policy frameworks and guidelines by which to further such objectives.

Finally, the model developed in this course can be expanded and adopted in other educational settings (e.g., PhD and master's student workshops, winter and summer schools, etc.). This initial experiment has provided valuable insights into the challenges as well as the possibilities of working within the setting of Genova. We hope that this work can inspire future endeavours to address these critical issues across a diverse range of urban, climate, and heritage contexts around the world. The more instances that can be developed, the more solutions can be identified to address the critical challenges currently facing cities. Putting students in a position to directly engage with these real-world problems as well as develop innovative solutions helps ensure that the next generation of practitioners and researchers will be adequately prepared to confront the challenges ahead.

Acknowledgments This research has been funded by the Politecnico di Milano, School of Architecture, Urban Planning and Construction Engineering (AUIC). All the produced materials are original contributions of the authors and their students. I.M. (the first author) has received funding from the Project "National Biodiversity Future Center—NBFC" funded under the National Recovery and Resilience Plan (NRRP), Mission 4 Component 2 Investment 1.4—Call for Tender No. 3138 of 16 December 2021, rectified by Decree No. 3175 of 18 December 2021 of the Italian Ministry of University and Research funded by the European Union—NextGenerationEU; Project code CN_00000033, Concession Decree No. 1034 of 17 June 2022 adopted by the Italian Ministry of University and Research, CUP, H43C22000530001.

Author contributions *I.M.* (Conceptualization, methodology, Data Curation, Content Validation, Resources, Writing original draft, reviewing, editing and Supervision); *Z.J.* (Writing—Original Draft Preparation, Writing—Review & Editing). *A.J & N.A.* both equally contributed to the Visualization and Data Curation. All authors agreed to the final submitted manuscript version.

References

Andreucci, M. B., Loder, A., Brown, M., & Brajkovi, J. (2021). Exploring challenges and opportunities of biophilic urban design: Evidence from research and experimentation. *Sustainability (Switzerland), 13*(Greening Cities), 4323. https://doi.org/10.3390/su13084323

Bockarjova, M., Botzen, W. J. W., Bulkeley, H. A., & Toxopeus, H. (2022). Estimating the social value of nature-based solutions in European cities. *Scientific Reports, 12*(1). https://doi.org/10.1038/s41598-022-23983-3

Bradley, S., & Mahmoud, I. H. (2024). Strategies for co-creation and co-governance in urban contexts: Building trust in local communities with limited social structures. *Urban Science, 8*(1), 9. https://doi.org/10.3390/urbansci8010009

Bulkeley, H. (2020). *Moving towards transformative change for biodiversity: Harnessing the potential of the post-2020 global biodiversity framework.* UK Centre for Ecology & Hydrology.

CBD. (2022). Kunming-Montreal Global biodiversity framework. In: *Draft decision submitted by the President*. Convention on Biological Diversity. https://www.cbd.int/conferences/2021-2022/cop-15/documentsasdocument:CBD/COP/15/L25

Convention on Biological Diversity. (2021). *FIRST DRAFT OF THE POST-2020 GLOBAL BIODIVERSITY FRAMEWORK*. https://www.cbd.int/doc/c/abb5/591f/2e46096d3f0330b08ce87a45/wg2020-03-03-en.pdf

Europa Nostra & ICOMOS. (2021). *European cultural heritage green paper: Putting Europe's shared heritage at the heart of the European Green Deal*. Europa Nostra. https://openarchive.icomos.org/id/eprint/2552/1/2021_European%20Heritage%20Green%20Paper_full%20paper.pdf

European Commission. (2023). *Guidelines for co-creation and co-governance of nature-based solutions-insights from EU-funded* projects (I. Ferreira, G. Lupp, & I. Mahmoud, Eds.). Publications Office of the European Union. https://doi.org/10.2777/157060

Galle, N. J., Nitoslawski, S. A., & Pilla, F. (2019). The internet of nature: How taking nature online can shape urban ecosystems. *Anthropocene Review, 6*(3), 279–287. https://doi.org/10.1177/2053019619877103

ICOMOS. (2019). *Climate Change and Cultural Heritage Working Group. The future of our pasts: Engaging cultural heritage in climate action*. ICOMOS. https://civvih.icomos.org/charters/2019-future-of-our-pasts-report/

ICOMOS. (2022). *Climate change adaptation*. ICOMOS. https://www.icomos.org/images/DOCUMENTS/Secretariat/2022/Climate_Action_Working_Group/CAWG-adaptation-english.pdf

IUCN (Ed.). (2020). *IUCN Global Standard for Nature-based Solutions: A user-friendly framework for the verification, design and scaling up of NbS* (1st ed.). IUCN, International Union for Conservation of Nature. https://doi.org/10.2305/IUCN.CH.2020.08.en

Jones, Z. M. (2022). Short-term gains and long-term challenges to learning from mega-event planning in the city of Genoa. *European Planning Studies, 30*(3), 437–456. https://doi.org/10.1080/09654313.2021.1959720

Kamjou, E., Scott, M., & Lennon, M. (2024). A bottom-up perspective on green infrastructure in informal settlements: Understanding nature's benefits through lived experiences. *Urban Forestry and Urban Greening, 94*. https://doi.org/10.1016/j.ufug.2024.128231

Kowarik, I., Fischer, L. K., Haase, D., Kabisch, N., Kleinschroth, F., Konijnendijk, C., Straka, T. M., & von Haaren, C. (2025). Promoting urban biodiversity for the benefit of people and nature. *Nature Reviews Biodiversity*. https://doi.org/10.1038/s44358-025-00035-y

Mahmoud, I. H. (2022). Public Spaces and Placemaking Approach: The Implementation Strategies. In: Placemaking for Green Urban Regeneration. The Urban Book Series. Springer, Cham. (pp 9-28). https://doi.org/10.1007/978-3-031-15408-9_2

Mahmoud, I. (2024). Nature-based solutions across different spatial urban scales: An analysis of three cases studies from Nice, Utrecht, and Milan. *Proceedings of the Institution of Civil Engineers: Urban Design and Planning*. https://doi.org/10.1680/jurdp.23.00063

Mahmoud, I., Dubois, G., Liquete, C., & Robuchon, M. (2025). Understanding collaborative governance of biodiversity-inclusive urban planning: Methodological approach and benchmarking results for urban nature plans in 10 European cities. *Urban Ecosystems, 28*(2), 17. https://doi.org/10.1007/s11252-024-01656-5

Paliaga, G., Luino, F., Turconi, L., & Faccini, F. (2018). Inventory of geo-hydrological phenomena in Genova municipality (NW Italy). *Journal of Maps, 15*(2), 28–37.

Roggema, R. (2020). *Nature-driven urbanism*. https://doi.org/10.1007/978-3-030-26717-9_1

UNEP. (2022). *Nature-based solutions for supporting sustainable development*. https://www.unep.org/topics/nature-action/nature-based-solutions/overview-nature-based-solutions

Viti, M. M., Gkimtsas, G., Liquete, C., Dubois, G., Borg, J., Dalla Costa, S., Teller, A., Hauser, R., & Robuchon, M. (2024). Introducing the progress monitoring tools of the EU Biodiversity Strategy for 2030. *Ecological Indicators, 164*, 112147. https://doi.org/10.1016/j.ecolind.2024.112147

Tracing Nature in an Iranian City: A Socio-spatial Analysis of Natural Spaces in Shiraz

Ayda Rastiemadabadi

Abstract Traditionally, cities maintained a closer relationship with nature due to their smaller size and lower density. However, with the onset of the industrial age, rapid population growth, the rise of metropolises, and shifts in urban planning aimed at accommodating expansion, nature became increasingly detached from daily urban life. Since then, the need to incorporate nature into cities has emerged not as a completely natural element but as an integral part of urban plans and programmes. Thus, understanding the relationship between contemporary cities and nature, along with an awareness of their historical roots, provides the foundation for making more informed and precise decisions in urban planning and design. This chapter takes a comprehensive approach to exploring the city's natural layer, aiming to better understand these urban spaces and their interaction with the city. Shiraz, an Iranian city with a long history, has been selected as the case study. Nature is an integral part of this city's identity, both structurally and culturally. This study employs a socio-spatial analysis to investigate the natural spaces' structure and dynamic characteristics. Additionally, their role in shaping the city's identity, structure, and culture is examined by tracing the evolution of various green spaces and natural resources embedded within Shiraz. The role of natural spaces in the chain of urban social life is also examined by categorising them and observing their daily life rhythms.

Existing research on urban green spaces often focuses solely on quantitative metrics such as location, distribution, and area or examines the quality of only a few selected cases. In contrast, this study considers nature as a holistic layer. Employing a socio-spatial methodology, it traces the historical layering of natural spaces within the city's evolution, analyses their socio-cultural relevance in shaping urban character and everyday life, and examines their structural and functional roles across different periods of development. Such insights can inform more deliberate and environmentally conscious urban planning strategies that lead to more sustainable

---

A. Rastiemadabadi (✉)
Department of Urbanism, Territory and Landscape, Polytechnic University of Catalonia, Barcelona, Spain
e-mail: ayda.rastiemadabadi@upc.edu

A. Cheshmehzangi et al. (eds.), *Designing Nature into Cities*, Urban Sustainability, https://doi.org/10.1007/978-981-95-9382-8_6

cities. Additionally, the methodology and findings of this study can be applied to investigate the role of green and natural spaces in shaping the character of other cities, allowing for comparative analyses that enhance our understanding of urban–nature relationships across various contexts.

Keywords Urban natural space · Socio-spatial analysis · Iranian city · Urban identity · Urban character · Urban evolution

1 Introduction

Cities have traditionally had a closer relationship with nature due to their smaller size and lower density. Early cities were closely connected to their natural surroundings (Tarr, 2010). They were often built in places that had natural advantages, like waterways and fertile land. Nature was used in early urban planning to improve living conditions and create a close connection between the shape of cities and natural features in traditional towns (Mumford, 1961). Generally, with their smaller, less dense urban forms, traditional cities allowed for a symbiotic relationship with nature, efficiently utilising local resources and adapting to climatic conditions (Giovagnorio & Chiri, 2016). However, with the onset of the industrial age, rapid population growth, and the rise of metropolitan areas, recent research confirms that, even today, higher human population density is associated with less tree cover, implying that as cities have grown denser, the integration of natural elements has frequently decreased (McDonald et al., 2023). Accordingly, integrating natural elements into urban planning is required to address current environmental concerns (Harms et al., 2024). Collaboration among urban planners, policymakers, and communities is critical for creating cities that promote ecological health, social well-being, and long-term sustainability. As a result, a comprehensive, integrated approach to urban design is required, balancing urban development demands with the preservation and enhancement of natural ecosystems (Ojo, 2024). Such integration can improve urban resilience and shift underlying worldviews to prioritise reciprocity with nature (Beatley, 2011).

Thus, understanding the relationship between contemporary cities and nature, along with an awareness of their historical roots, provides the foundation for making more informed and precise decisions in urban planning and design. Therefore, this discernment provides a suitable context for making more informed decisions in urban planning and design. In this regard, first, theoretical concepts related to urban nature and its evolution, especially in the context of Iranian cities, are examined. Then, the city of Shiraz is introduced as a case study, and its historical background is explained in relation to natural elements. Using socio-spatial analysis, how nature affects the city's identity, physical structure, and daily life across different periods of urban development is examined. In the final sections, the results obtained are analysed and interpreted, and their implications for sustainable urban development are discussed, along with suggestions for future research in this field.

This chapter aims to investigate the role of natural spaces in shaping Shiraz's urban identity, structure, and socio-cultural life through a socio-spatial perspective by addressing the following research questions:

1. How have natural elements historically shaped Shiraz's identity, structure, and daily life across different periods of urban development?
2. In what ways do natural spaces contribute to the socio-cultural character of the city today?
3. How can a socio-spatial methodology provide a framework for understanding the relationship between nature and urban identity in Shiraz and comparable contexts?

1.1 Nature and/in Cities

As a multidimensional concept, urban nature includes natural and semi-natural elements within urban environments that play an important role in shaping human life and maintaining ecological balance (Hoyle, 2020; Kühne, 2012; Lynch, 1960; Visintin et al., 2024). Beatley (2011), in his book Biophilic Cities, provides a comprehensive definition of urban nature, describing it as any green space, body of water, or natural habitat located within the urban context. These elements include parks, gardens, street trees, rivers, community gardens, and green roofs. Urban nature is not limited to natural landscapes but encompasses all elements that connect people with nature and help improve citizens' quality of life (Tzoulas et al., 2007). However, the precise definition of urban nature remains challenging, as the concept is influenced by various factors, including culture, history, and geographical conditions (Jiang et al., 2015). For example, urban nature may include untouched natural elements in some cities, while in more industrialised cities, it may take the form of artificial green spaces or green roofs. This diversity in definitions highlights that urban nature is flexible and shaped by local conditions. On the other hand, although the separation of nature and the city is common in language, from both environmental and social perspectives, the city is an inseparable part of nature, and nature is deeply intertwined with the social life of cities (Benton-Short & Short, 2013). Several studies have examined the different roles of urban nature, such as:

Ecological functions: An essential part of urban ecological systems is urban nature. Studies show it maintains biodiversity, controls microclimates, and reduces the adverse environmental effects of crowded populated areas (Grimm et al., 2008). Integrating natural elements within urban design can also enhance resilience and sustainability (Pedersen Zari et al., 2022), highlighting the need to treat ecological layers as an integral part of urban planning.

Public health and well-being: It has been argued that urban nature can help restore mental focus and reduce stress, thereby enhancing citizens' quality of life (Kaplan & Kaplan, 1989). Gehl and Rogers (2010) argue that natural elements are not merely aesthetic but essential to maintaining urban populations' mental, physical, and social vitality.

Cultural and symbolic roles: Urban nature also carries deep cultural and historical significance. In this regard, case studies from different cities have shown that urban nature can play various roles, including serving as cultural symbols, aesthetic elements, or components of urban ecological networks. For example, Nazarian (2002), in his book Urban Geography of Iran, examined the role of urban nature in shaping urban identity and demonstrated that these elements can act as cultural and historical symbols of the city. Similarly, Hoseini et al. (2012), in a study on urban green spaces in Shiraz, found that urban nature in this city functions not only as an aesthetic element but also as a part of local culture and traditions.

These studies indicate that understanding the concept of urban nature requires the integration of multiple perspectives—ecological, cultural, and social. This integrated approach can support the development of urban strategies that respond to citizens' functional needs and help maintain a balance between city and nature.

1.2 Iranian Cities and Nature

Iranian cities have long been intertwined with nature, with natural elements such as water, trees, and mountains playing a central role in shaping their identity and structure (Kheirabadi, 2000; Partovi & Motamedi, 2023). Traditional Iranian cities maintained their security, vitality, and longevity by integrating natural elements such as rivers and foothills into their urban fabric. At the same time, they fostered urban growth through environmentally conscious engineering (Habibi & Hourcade, 2005). For example, the qanat system provided a sustainable method for delivering water to cities (Alemohammad & Gharari, 2010). Water resources were also used effectively to fertilise the soil, and gardens of various sizes were established within and around urban areas (Mansouri & Haidartaj, 2010). These green spaces play a crucial role in moderating the urban climate (Yazdi & Karimian, 2021). Since ancient times, Iranian gardens have exemplified this harmonious design approach (Mansouri, 2005).

Additionally, nature was integrated into daily life at multiple scales in traditional Iranian cities—for instance, through the architectural use of water in house and garden ponds—reflecting both a respect for nature and a conscious effort to incorporate it into everyday living. This approach continued into the Islamic period, with the architectural integration of mosques, water sources, and Iranian gardens serving as a clear expression of this harmony (Pakzad, 2011; Rastiemadabadi et al., 2025).

In the historical cities of Iran, natural elements were valued not only as material resources but also as spiritual and cultural symbols (Moussavi-Aghdam, 2012; Nasr, 1997). For example, specific natural sites were used for religious and social ceremonies in the past, though urbanisation gradually affected these traditions (Ehlers & Floor, 1993). While Iranian cities have long strived to balance development with respect for nature, population pressures and industrialisation have disrupted this equilibrium (Madanipour, 2003). The unsustainable expansion of cities in recent decades has led to decreased biodiversity and the destruction of natural landscapes on the urban periphery (UN-Habitat, 2010). These challenges underscore the need for future urban planning to draw on historical lessons in order to restore harmony between the city and nature (Soltanzadeh, 2011).

In comparison with the Western literature on urban nature, which frequently discusses ecosystem services, public health benefits, and recreational value (Beatley, 2011; McDonald et al., 2023), Iranian traditions have historically embedded natural elements within the cultural, spiritual, and spatial identity of cities (Kheirabadi, 2000; Mansouri, 2005). In Western contexts, approaches such as Nature-Based Solutions (NBS) and biophilic design are largely discussed as strategies for climate adaptation, resilience, and well-being (Ben Hassen & Hageer, 2023; Cook et al., 2025; Tzoulas et al., 2007; Zarei & Shahab, 2025), whereas in Iranian cities, gardens and water networks have functioned not only as ecological infrastructure but also as symbolic and social representations of spatial and cultural order (Habibi & Hourcade, 2005; Nasr, 1997). This distinction provides a more transparent theoretical framework for analysing how Shiraz's natural layers operate as both environmental systems and cultural–spatial anchors of urban identity.

2 Materials and Methods

This research, using a qualitative approach and socio-spatial analysis, examines the relationship between nature and urban structure, identity, and daily life in Shiraz. The research time frame is defined based on the study of historical developments in the city's formation and development period. This temporal classification allows for the analysis of the evolution of the presence and role of nature in the context of the city's social, cultural, and physical developments. The research data were collected through a combination of documentary and historical studies, analysis of old and new maps, visual documentation, and field observations.

To trace the presence of nature in the city, two main dimensions were examined: space and mass (filled texture). In the spatial dimension, the natural layers of the city, including mountains (brown), water resources (blue), and vegetation (green), were analysed as the main elements. These layers were examined in relation to the formation of the city's physical structure and its expansion. In the mass analysis section, the urban fabric of Shiraz was divided into typologies, each reflecting density, architectural style, and relationship to nature. Also, the city's living spaces were classified into three general categories: private and semi-private, semi-public, and public spaces, in order to study the presence and integration of nature into the daily lives of residents. This combined and layered approach provides a comprehensive perspective for understanding the place of nature in the spatial, cultural, and social structure of Shiraz (see Fig. 1).

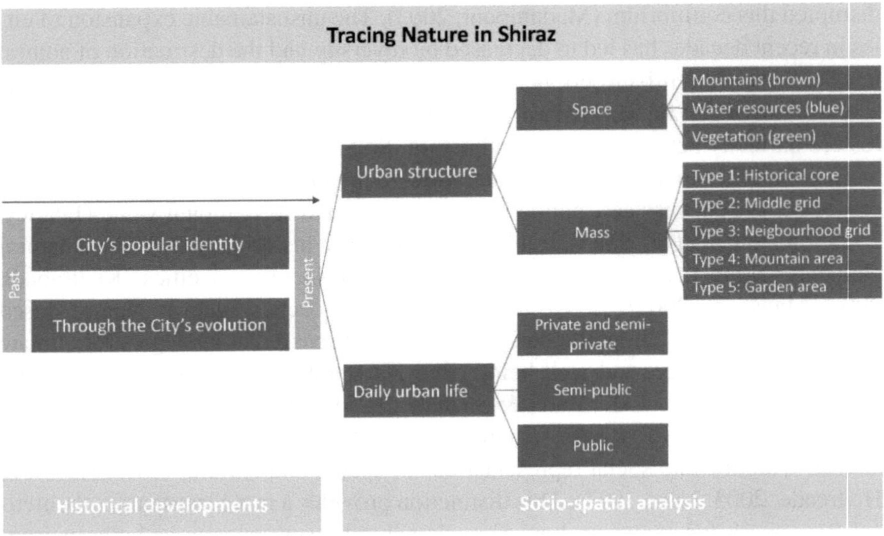

Fig. 1 Methodological framework of the chapter (by the author)

3 Case Study: Shiraz

Shiraz is the fifth most populous city in Iran and the capital of Fars Province. It covers an area of approximately 250 square kilometres and is located in southwestern Iran at an average altitude of 1540 m above sea level. In 2016, the city had a population of 1,869,001. Shiraz has served as the capital of Fars Province since the Islamic conquest, succeeding Estakhr during the Sasanian period and Persepolis during the Achaemenid era. It was situated at the crossroads of several important routes in ancient Iran, at the heart of Fars Province—one of the most significant regions of the time. This role as a central hub has contributed to its distinctive urban structures (Ardeshiri & Memarian, 2010). The reason for selecting this city is that its identity is deeply intertwined with nature. Various natural layers have played a role in Shiraz's urban and cultural formation. Elements such as the river running through the city, historical Iranian gardens in different districts, many private gardens embedded in the urban fabric, and the surrounding foothills have all shaped the city's overall structure and clearly illustrate its longstanding coexistence with nature (Rastiemadabadi, 2015).

Shiraz demonstrates a particularly close integration of natural elements within its urban fabric compared to other Iranian cities and especially major metropolises. The rapid expansion of Tehran has largely overshadowed its historical green spaces, leaving nature fragmented and increasingly dependent on recent urban planning interventions. Although historically renowned for the Chaharbagh axis (green axis) and the Zayandeh Rud (main river), Isfahan reflects a more formal and planned relationship between nature and urban space. Shaped by industrial growth and transport-oriented development, Tabriz has partially lost its organic connection to natural layers, requiring contemporary planning efforts to restore this relationship. Despite its significant religious axis and pilgrimage-centred development, Mashhad has frequently reduced natural spaces to peripheral roles within the city. On the other hand, Shiraz demonstrates a multi-layered and natural blending of gardens, foothills, and rivers that pass through public and private spaces, establishing a distinctive socio-spatial conversation between urban and natural elements. This combination of historical continuity and organic interaction with nature makes Shiraz stand out among Iranian metropolises as a particularly valuable case for exploring the cultural and spatial role of natural layers in urban identity.

4 Tracing Nature in Shiraz

4.1 Nature and the City's Popular Identity

The identity of a city is formed over time and serves as the essence and reflection of what has occurred within the city and among its people. Human experience and behaviour are shaped through a network of memories and identities closely linked to the environment. Therefore, a city's identity can be examined by understanding a society's socio-environmental values and the dynamic relationship between humans and their surrounding environment (Cheshmehzangi & Heat, 2012). In this section, to trace the relationship between the city and nature, the first step is to examine what Shiraz is known for, its general identity, and to determine the role of nature in shaping the city's character.

Shiraz has long been celebrated for its gardens, flowers, wine, and poetry, earning titles such as the Seat of Government, the Abode of Knowledge, and the Tower of Saints. Its natural setting, with a temperate climate, abundant water sources, and fertile soil, has fostered the development of lush gardens characterised by cypress, orange, and pine trees, as well as seasonal flowers like narcissus. Historical visitors such as Tavernier and Phillips (1678), Herbert and Butler (2012), and Christie (2011) praised Shiraz's greenery and paradisiacal atmosphere.

In addition to its natural beauty, Shiraz was historically renowned for its wine, which was once considered the finest in the Middle East. Although wine production is now banned, vineyards still shape the cultural landscape of the city. Shiraz's intellectual and spiritual heritage is equally significant. Known as a city of poets and saints, many of whom lived in the pleasure gardens of Shiraz or the serene mountains surrounding it, the city has often been depicted as a paradise in its poetry, writings, and philosophical works. The rulers of Shiraz embraced this green culture and established gardens in the Persian garden style, symbolising paradise through the combination of blue and green. These gardens remain a significant part of the city's urban heritage. Additionally, the city's scholarly legacy is anchored by the Khan School and continues today through institutions such as Shiraz University, upholding its title as the Abode of Knowledge. The famous Khan School and other historical institutions were also designed with the Persian garden tradition.

Today, the city retains this identity through four main types of green spaces: historic gardens, urban parks, extensive private fruit gardens such as those in Gasrodasht area, and landscaped streets. Nature is deeply tied to the character of Shiraz, and this enduring identity continues to foster a close and harmonious relationship between people and nature, even amid the city's ongoing urban development.

4.2 Nature and the Urban Evolution

Like other Iranian cities, Shiraz follows a specific pattern of change and development, which can be divided into four distinct periods. To better understand the city's relationship with nature during each period, green spaces, water resources, and major mountains are examined in relation to the city's structure (see Fig. 2).

Shiraz's traditional core (Before 1921): In the traditional era, the city was enclosed within walls, and most houses featured a central courtyard with a pond and a garden. These features were generally proportional to the size of the house and the homeowner's financial means. Gardens and agricultural lands were primarily located outside the city walls, except for some royal and private gardens within the walls or at the city's entrance.

Shiraz's first modern period (1921–1941): During this time, the city began to expand by developing new modern streets, including a main boulevard. Following this initial expansion beyond the city walls, only those with sufficient financial means and reliable access to water and security moved outside the traditional core. This led to the emergence of private garden houses surrounding the historical city.

Shiraz's second modern period (1941–1979): In this period, the construction of streets gave rise to new neighbourhoods, and a new architectural style in residential buildings was introduced to Iranian urban planning for the first time. Nevertheless, gardens and ponds remained common features in residential courtyards. Some government-owned gardens were also opened to the public for the first time during this era.

Shiraz after the Islamic Revolution (since 1979): The most significant urban changes occurred during this period, marked by a population boom and rapid urban expansion. Apartment buildings became widespread, and the increasing demand for land led to the disappearance of much of the agricultural land surrounding the city.

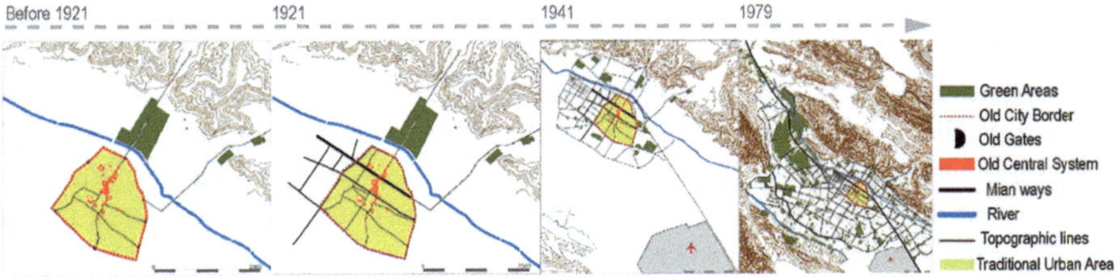

Fig. 2 Evolution of Shiraz and the distribution of main natural resources in relation to urban development over time (by the author)

Many nearby villages, along with their water sources, fruit orchards, and farmland, were incorporated into the city's boundaries (see Fig. 3).

However, a remarkable aspect of Shiraz, attributed to its nature-friendly identity and urban culture, is that the pace of high-rise development and the loss of natural or agricultural land occurred more slowly than in other Iranian cities. Even today, many residents prefer to live in houses with courtyards or to own private gardens within or outside the city, giving Shiraz a distinct character compared to other Iranian metropolises.

4.3 *Nature and the Urban Structure*

As previously explained in the methodology, to analyse and trace nature's presence in Shiraz's structure, the city has been examined through two main categories: space and mass. Regarding the "space" aspect, three main natural components were identified for the analysis of the natural layers of Shiraz: brown for mountains, blue for various water sources, and green for natural and artificial vegetation (see Fig. 4). These components were selected from the many layers that make up the city. By examining these elements together, several significant observations emerge.

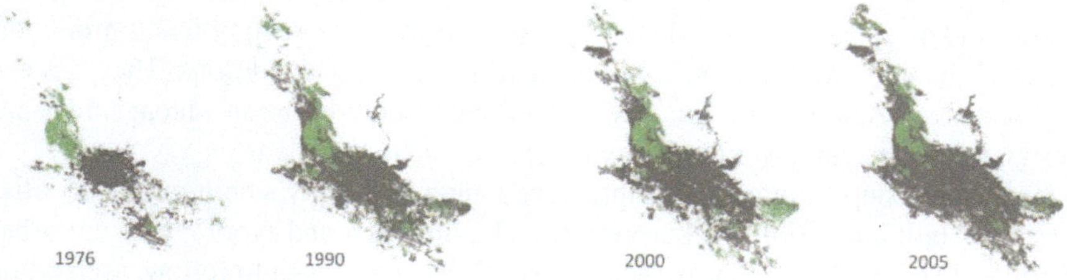

Fig. 3 Urban growth of Shiraz after the Islamic Revolution, showing expansion from 1976 to 2005 (re-illustrated by the author)

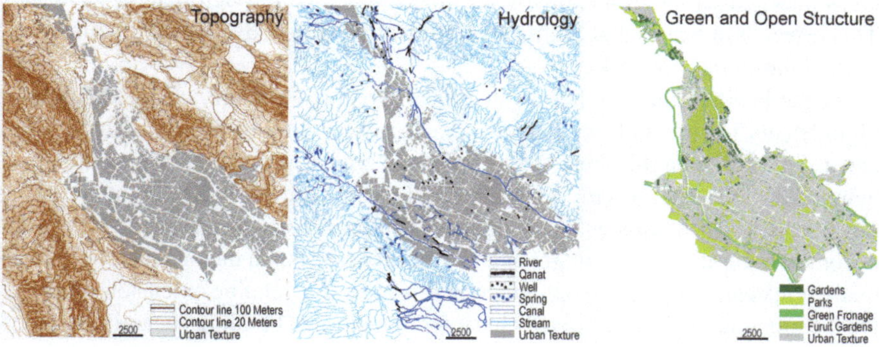

Fig. 4 The three main natural layers of Shiraz's urban structure: mountains (brown), water resources (blue), and vegetation (green) (by the author)

The mountains surrounding Shiraz form a natural boundary around the city. In the past, this topographic position played an important defensive role. Today, it significantly influences air circulation and pollution control. This topographic feature is the most important factor shaping the form and structure of the city. Furthermore, the mountains near the city are not particularly high, providing accessible slopes that people can easily climb to gain a panoramic view of the city, and this activity has become a part of local culture.

Shiraz has a variety of water sources, including rainfall, groundwater, seasonal streams, rivers, springs, aqueducts, and wells. Today, the city's main drinking-water sources are wells and groundwater. Water canals used for gardens and farms within the city significantly impact the microclimate and occasionally create picturesque scenes that draw public attention. In Shiraz, water remains visible in many urban settings: within houses and gardens, parks, tomb complexes, traffic squares, public decorations, artificial lakes, and even medians between streets as landscaped refuges.

Due to its unique climate and traditional culture of gardening, Shiraz possesses a rich vegetation cover. Although the city's rapid expansion in recent years has led to the destruction of many gardens and farms around its periphery, substantial green areas within the city still contribute to its distinctive character. Additionally, protected vegetative zones surrounding the city serve important ecological functions and recreational spaces for residents.

This analysis demonstrates that natural elements play a vital role in shaping Shiraz's identity and functionality. These features have been significant throughout the city's history and continue to influence its present-day conditions. They serve as key factors in determining the city's form and structure, urban management and design strategies, and overall quality of life for its citizens.

In the next step, nature's footprints were sought in the city's built mass—its filled spaces or buildings. For this purpose, the classification and typology of the urban fabric were employed. To make the analysis clearer and easier to follow, each typology was given a short descriptive name (e.g., Early Modern Grid). The entire city plan was then mapped and summarised according to these types to show their locations and spatial relationships. To reduce subjectivity, typology boundaries were cross-checked with multiple data sources and validated through field observations. This research indicates that Shiraz predominantly exhibits five types of built texture or granularity (see Table 1).

At the heart of the city lies its historical, organic structure, characterised by a high-density texture and tiny parcels. One- or two-storey buildings contribute to a consistent skyline in this area. Open spaces are typically located at the centre of old neighbourhoods and usually contain small shops and public buildings such as mosques and open squares. Although the compact texture and narrow alleys are devoid of natural elements or green and blue landscapes, on the one hand, the handmade bricks and organic layout integrate this area as a genuine extension of nature. On the other hand, the introverted character of historical buildings beautifully incorporates the natural landscape within their boundaries. The courtyards, reminiscent of paradise, accommodate green trees and blue water ponds on various scales. These

Table 1 Summary of Shiraz's five main urban fabric typologies with key characteristics such as density, greenery, building form, and spatial relationships (by the author)

Typology	Density	Greenery/ nature	Building form and height	Street pattern	Key characteristics
1. Historical Core	High	Courtyards with trees; small ponds	1–2 storeys; small parcels	Organic, narrow alleys	Dense texture; introverted houses integrating nature inside plots; cultural heritage
2. Early Modern Expansion	Medium–High	Tree-lined boulevards; few large parks	Tall buildings; large parcels	Grid pattern, wide boulevards	Hosts most facilities and services; modernisation phase with limited but structured greenery
3. Contemporary Neighbourhoods	Medium	Small-medium parks; scattered greenery	Mixed heights; moderate parcels	Grid-like; planned	Most common fabric; suburban expansion with integrated but moderate green areas
4. Hillside Neighbourhoods	Medium	Adapted to topography; natural slopes	Medium density; variable heights	Curved streets following slope	Mountain-adapted design; combines planned layouts with natural terrain
5. Garden Areas	Low	Extensive gardens; agricultural remnants	Villas and rural houses; mostly ≤ 3 storeys	Organic in rural sections	Functions as city's "lungs"; combines luxury and traditional garden structures

elements not only enhance the aesthetic quality but also contribute to the functionality of the space: the gardens help cool the interiors, while the central pond serves as a water reservoir.

The second type of texture corresponds to the first area that experienced urban growth during the modern period, marked by the demolition of historical walls and the introduction of wide boulevards and a grid street pattern. Due to its central location, this area hosts most of the city's facilities, services, offices, and commercial centres. Consequently, it contains large parcels and tall buildings, maximising land value. From its inception to the present, this area has been characterised by rows of trees along and between the streets, although large green spaces are rare. Despite being the most bustling part of the city, walking along its sidewalks and alleys remains a pleasant experience for citizens.

This third type represents the most common urban fabric in contemporary Shiraz. It includes various neighbourhoods with differing layouts, characteristics, and basic amenities. Typically, these neighbourhoods feature a range of building heights, a grid-like street pattern, and a moderate average density. Small and medium-sized parks contribute to the modern neighbourhood design in Shiraz.

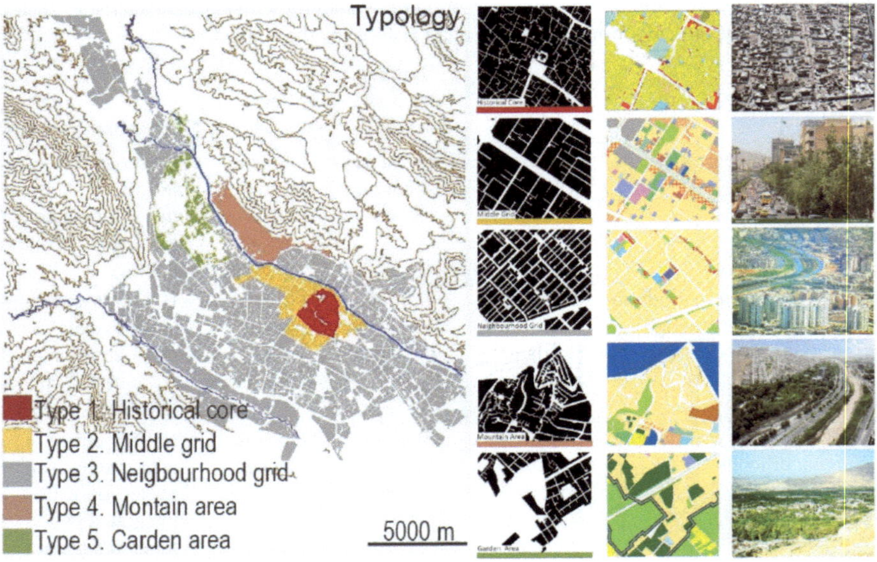

Fig. 5 Spatial distribution and visual representation of Shiraz's five main urban fabric typologies: Historical Core, Middle Grid, Neighbourhood Grid, Mountain Area, and Garden Area, showing their locations, patterns, and characteristic features (by the author)

Natural elements significantly shape the remaining two types of urban structure and texture.

The hillside neighbourhoods share many qualities with other planned neighbourhoods. However, their topographical context places them in a distinct category. This area generally consists of medium-sized and moderate-density parcels with buildings of varying heights. The street network is fully adapted to the mountain's natural slope.

The garden area, which covers approximately 1300 ha, mainly combines luxury villas with remnants of rural housing, all surrounded by gardens. As such, it exhibits low density, large parcels, and a building height of no more than three storeys. In the rural segments, parcels are smaller and arranged in an organic pattern. This area functions as the city's lungs and endows Shiraz with a distinctive character (see Fig. 5).

This analysis shows that despite the differences in density, form, and historical periods across Shiraz's urban contexts, the imprint of nature has remained deeply embedded within all of these texture patterns. Whether in the introverted courtyards of the historical core, the tree-lined boulevards of modern development, or the sprawling gardens and mountainside neighbourhoods, natural elements continue to play a fundamental role in shaping the identity and experience of urban life. These findings emphasise the enduring presence of nature in Shiraz's urban development process and demonstrate how natural elements have adapted to and persisted through the city's morphological changes.

4.4 *Nature in the Daily Urban Life*

In order to study nature in the contemporary life of Shiraz, living spaces are first divided into three main categories according to their hierarchy: private and semi-private (including private houses, gardens, and courtyards of residential complexes), semi-public (such as public buildings, cultural centres, cafes and restaurants, historic houses and gardens, and garden-tombs of poets and philosophers), and public (streets, parks, and the natural surroundings of the city). This classification highlights an essential aspect of the connection between nature and living spaces in Shiraz.

In examining private spaces, it is evident that Shiraz still favours independent houses with courtyards and gardens. As a result, the city is characterised by a relatively low height and low-density urban form compared to other Iranian metropolises. The people's enduring interest in spending time in gardens has preserved many private gardens within the city and encouraged the development of extensive private garden areas around it.

Within the private and semi-private categories, old gardens and historical houses that have been opened to the public offer pleasant spaces for leisure and social interaction. The abundance of such sites has helped Shiraz maintain its longstanding reputation as the "city of gardens". Moreover, cafes and restaurants integrated into natural settings are highly popular in Shiraz. In addition, the tombs of renowned poets and philosophers have been mainly transformed into gardens, contributing to the city's rich collection of natural spaces (see Fig. 6).

Finally, public spaces such as streets are often lined with trees. In some cases, the sides of highways have been converted into linear parks—the most notable of which follows the course of the city's river, flanked by a park stretching more than 3 km.

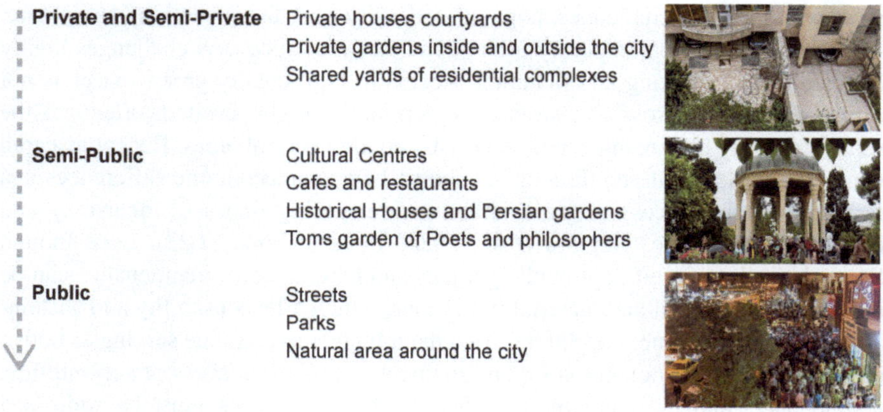

Fig. 6 The flow of life in Shiraz through the chain of its urban spaces (Top: Private house courtyard, ISNA News Agency, by Azam Dadashi. Middle: Tomb of Hafez, A poet, by Mohammadreza Abdollahi. Bottom: Zand Street in Shiraz, Mehr News Agency, by Amin Berenjkar).

The surrounding hills and mountains are also popular recreational destinations and have been designated as protected areas with the development of walking paths.

It remains unclear whether Shiraz's natural landscape and setting have fostered a deep-rooted cultural appreciation for nature from ancient times to the present, or whether the longstanding interest of its residents has led to the preservation and expansion of natural elements throughout the city. Nonetheless, it is evident that nature holds a special place in the culture and daily life of the people in Shiraz today.

5 Discussion

The results of this study indicate that nature in the city of Shiraz is more than a decorative or merely functional element; it plays a fundamental role in shaping urban identity, spatial structure, and public culture. The continuous and structured presence of elements such as gardens, canals, rivers, and surrounding mountains has persisted not only in the historical context but also within the contemporary urban fabric, setting Shiraz apart from other Iranian metropolises. This finding aligns with previous studies that highlight the role of natural elements in fostering a sense of place and urban identity (Cheshmehzangi & Heat, 2012; Kaplan & Kaplan, 1989; Nazarian, 2002). Unlike many modern cities where urban expansion has diminished the human–nature connection, in Shiraz, nature continues to be an integral part of daily life.

These findings are also consistent with international theories and studies that underscore the importance of liveable cities and biophilic urbanism (Beatley, 2011; McDonald et al., 2023). Similarly, other studies on green spaces in Shiraz affirm the deep cultural and social significance of nature in the city (Hoseini et al., 2012). This insight offers a strategic model for other cities in Iran and beyond that are grappling with environmental challenges, population density, and disconnection from nature. At the same time, contemporary urban dynamics introduce new challenges highly relevant to placemaking and planning. Access to high-quality green spaces is not evenly distributed across all districts, and peripheral neighbourhoods often lack the same proximity and maintenance standards found in central areas. Recent research conducted in Istanbul and Tehran has shown how socioeconomic differences can result in unequal access to green infrastructure, posing issues of inclusivity and environmental justice (Naghibi et al., 2024; Yazar & York, 2023). Even though Shiraz has a lot of well-kept public gardens and natural areas, maintenance can be restricted by financial and material limitations, which affects usability and sustainability over a long time. For Shiraz's rich natural layer to continue serving as both a cultural asset and a foundation for environmental resilience, issues of accessibility, maintenance, and the equitable distribution of green spaces must be addressed (Mohamed & Kronenberg, 2025).

6 Conclusion

By emphasising the critical role of natural layers in the structure and meaning of the city, this study opens new directions for reimagining sustainable urban development. The main conclusion is that nature in Shiraz embodies both a lasting heritage and a living future, offering valuable lessons for urban policymakers seeking to reinforce local identity, enhance the quality of life, and restore meaningful human–nature connections.

These insights provide a strategic reference point for other Iranian cities facing rapid urbanisation and environmental stress. This study clearly shows that reconnecting the city with natural layers is not only an ecological or aesthetic necessity but also a socio-cultural one that can fill the gap between historical continuity and modern urban demands.

Nevertheless, the study acknowledges several limitations, including the absence of a quantitative ecological assessment and limited direct community participation. Future research could address these gaps through participatory mapping, in-depth interviews, and mixed-method evaluations combining socio-spatial and ecological data to deepen understanding of urban–nature relationships in Shiraz and beyond.

References

Alemohammad, H., & Gharari, S. (2010). *Qanat: An ancient invention for water management in Iran*. Water History Conference, Delft, The Netherlands.

Ardeshiri, M., & Memarian, G. (2010). *Urban identity of Shiraz*. Navid Press.

Beatley, T. (2011). Biophilic cities: Integrating nature into urban design and planning. Island Press. https://books.google.com.tr/books?id=H9Y4z68WSgUC

Ben Hassen, T., & Hageer, Y. (2023). Urban climate resilience in MENA region: Opportunities and challenges of nature-based solutions. In W. Leal Filho, G. J. Nagy, & D. Y. Ayal (Eds.), *Handbook of nature-based solutions to mitigation and adaptation to climate change* (pp. 1–23). Springer International. https://doi.org/10.1007/978-3-030-98067-2_161-1

Benton-Short, L., & Short, J. R. (2013). *Cities and nature* (2nd ed.). Routledge. https://doi.org/10.4324/9780203103500

Cheshmehzangi, A., & Heat, T. (2012). Urban identities: Influences on socio-environmental values and spatial inter-relations. *Procedia - Social and Behavioral Sciences, 36*, 253–264. https://doi.org/10.1016/j.sbspro.2012.03.028

Christie, A. (2011). *The house at Shiraz: An Agatha Christie short story*. HarperCollins. https://books.google.com.tr/books?id=rp_FmXOLAUwC

Cook, E., Kim, Y., Grimm, N., McPhearson, T., Anderson, P., Bulkeley, H., Diep, L., Morató, J., & Zhou, W. (2025). Nature-based solutions for urban sustainability. *Proceedings of the National Academy of Sciences, 122*. https://doi.org/10.1073/pnas.2315909122

Ehlers, E., & Floor, W. (1993). Urban change in Iran, 1920-1941. *Iranian Studies, 26*(3/4), 251–275. http://www.jstor.org/stable/4310857

Gehl, J., & Rogers, R. (2010). *Cities for people*. Island Press.

Giovagnorio, I., & Chiri, G. (2016). *The environmental dimension of urban design: A point of view*. https://doi.org/10.5772/62883

Grimm, N. B., Faeth, S. H., Golubiewski, N. E., Redman, C. L., Wu, J., Bai, X., & Briggs, J. M. (2008). Global change and the ecology of cities. *Science, 319*(5864), 756–760. https://doi.org/10.1126/science.1150195

Habibi, M., & Hourcade, B. (2005). *Atlas of Tehran metropolis: Land & people* (Vol. 1). Shirkat-i Pardāzish va Barnāmah'rīzī-i Shahrī.

Harms, P., Hofer, M., & Artmann, M. (2024). Planning cities with nature for sustainability transformations — a systematic review. *Urban Transformations, 6*(1), 9. https://doi.org/10.1186/s42854-024-00066-2

Herbert, T., & Butler, J. A. (2012). *Travels in Africa, Persia, and Asia the great: Some years travels into Africa and Asia the great, especially describing the famous empires of Persia and Hindustan, as also divers other kingdoms in the Oriental Indies, 1627-30, the 1677 version*. ACMRS Arizona Center for Medieval and Renaissance Studies.

Hoseini, S. A., Veisi, R., & Ahmadi, S. (2012). Study and analyze of Shiraz green space. *The Journal Urban Ecological Research, 3*(5), 51–70. https://journals.pnu.ac.ir/article_1206_9ee37c97fbfcafc63817fef5ea89fd12.pdf

Hoyle, H. (2020). What is urban nature and how do we perceive it? In *Naturally challenged: Contested perceptions and practices in urban green spaces* (pp. 9–36). Springer. https://doi.org/10.1007/978-3-030-44480-8_2

Jiang, B., Zhang, T., & Sullivan, W. C. (2015). Healthy cities: Mechanisms and research questions regarding the impacts of urban green landscapes on public health and well-being [Report]. *Landscape Architecture Frontiers, 3*, 24+. https://link.gale.com/apps/doc/A453915481/AONE?u=anon~69f38ff8&sid=googleScholar&xid=22e9430e

Kaplan, R. S.-W., & Kaplan, S. (1989). *The experience of nature: A psychological perspective*. Cambridge University Press.

Kheirabadi, M. (2000). *Iranian cities: Formation and development*. Syracuse University Press.

Kühne, O. (2012). Urban nature between modern and postmodern aesthetics: Reflections based on the social constructivist approach. *Quaestiones Geographicae, 31*(2), 61–70. https://doi.org/10.2478/v10117-012-0019-3

Lynch, K. (1960). *The image of the city / Kevin Lynch*. MIT Press.

Madanipour, A. (2003). Modernization and everyday life: Urban and rural change in Iran. In A. Mohammidi (Ed.), *Iran Encountering Globalization*. Routledge. https://doi.org/10.4324/9780203824085

Mansouri, A., & Haidartaj, V. (2010). Hypotheses of Char-Bagh pattern in Persian gardens in critic. *Bagh-e-Nazar, 6*(12), 17–30.

Mansouri, S.-A. (2005). An introduction to the aesthetics of Iranian Garden. *The Monthly Scientific Journal of Bagh-e Nazar, 2*(3), 58–63. https://www.bagh-sj.com/article_1504_9e89b57f8dff54c7beb5b9006b9e0faf.pdf

McDonald, R., Aronson, M., Beatley, T., Beller, E., Bazo, M., Grossinger, R., Jessup, K., Mansur, A., Puppim de Oliveira, J., Panlasigui, S., Burg, J., Pevzner, N., Shanahan, D., Stoneburner, L., Rudd, A., & Spotswood, E. (2023). Denser and greener cities: Green interventions to achieve both urban density and nature. *People and Nature, 5*. https://doi.org/10.1002/pan3.10423

Mohamed, A. A., & Kronenberg, J. (2025). Users' experiences of park accessibility and attractiveness based on online review analytics. *Scientific Reports, 15*(1), 4268. https://doi.org/10.1038/s41598-025-88500-8

Moussavi-Aghdam, C. (2012). BUILDING IRAN: MODERNISM, ARCHITECTURE, AND NATIONAL HERITAGE UNDER THE PAHLAVI MONARCHS, Talinn Grigor. *The Arab Studies Journal, 20*(1), 204–208. http://www.jstor.org/stable/23265849

Mumford, L. (1961). *The city in history: Its origins, its transformations, and its prospects*. Harcourt, Brace & Co..

Naghibi, M., Farrokhi, A., & Faizi, M. (2024). Small urban green spaces: Insights into perception, preference, and psychological well-being in a densely populated areas of Tehran, Iran. *Environmental Health Insights, 18*, 11786302241248314. https://doi.org/10.1177/11786302241248314

Nasr, S. H. (1997). *Man and nature: The spiritual crisis in modern man*. ABC International Group. https://books.google.co.id/books?id=QAhYNAAACAAJ

Nazarian, A. (2002). *Urban geography of Iran*. Payam Noor University Press.

Ojo, D. (2024). Urban design and planning: Integrating nature and the built environment. In S. A. Neves Lousada & M. Fernandes (Eds.), *Landscape architecture and design - sustainability and management*. IntechOpen. https://doi.org/10.5772/intechopen.1006805

Pakzad, J. (2011). *History of Iranian city: From beginning to Qajar era*. Armanshahr Press.

Partovi, P., & Motamedi, R. (2023). A conceptual framework for water sensitive urban planning and design: Towards developing a fundamental tool to guide and assess policy making. *Journal of Architecture and Urban Planning, 15*(38), 63–79. https://doi.org/10.30480/aup.2022.3216.1673

Pedersen Zari, M., MacKinnon, M., Varshney, K., & Bakshi, N. (2022). Regenerative living cities and the urban climate–biodiversity–wellbeing nexus. *Nature Climate Change, 12*(7), 601–604. https://doi.org/10.1038/s41558-022-01390-w

Rastiemadabadi, A. (2015). *Contemporary Shiraz character: Parallel study on the social and morphological sides of urban spaces*. Universitat Politècnica de Catalunya, Barcelona. http://hdl.handle.net/2117/85073

Rastiemadabadi, A., Sabaté Bel, J., & Pesoa Marcilla, M. (2025). Traditional urban forms in contemporary cities: Morphological analysis of the traditional Iranian bazaars in present-day cities. *Open House International, 50*(2), 349–368. https://doi.org/10.1108/OHI-12-2023-0300

Soltanzadeh, H. (2011). *Brief HISTORY OF THE CITY AND URBANIZation in Iran*. Chahartagh Publications [in Persian].

Tarr, J. (2010). The city as an artifact of technology and the environment. In M. R. a. S. H. Cutcliffe (Ed.), *The illusory boundary: Environment and technology in history* (pp. 145–170). University of Virginia Press.

Tavernier, J.-B., & Phillips, J. (1678). *The six voyages of John Baptista Tavernier*. Printed for R.L. and M.P. https://www.biodiversitylibrary.org/item/148595

Tzoulas, K., Korpela, K., Venn, S., Yli-Pelkonen, V., Kaźmierczak, A., Niemela, J., & James, P. (2007). Promoting ecosystem and human health in urban areas using Green Infrastructure: A literature review. *Landscape and Urban Planning, 81*(3), 167–178. https://doi.org/10.1016/j.landurbplan.2007.02.001

UN-Habitat. (2010). *State of the world's cities 2010/2011 –cities for all: Bridging the urban divide*. UN-Habitat.

Visintin, C., Garrard, G., Weisser, W., Baracco, M., Hobbs, R., & Bekessy, S. (2024). Designing cities for everyday nature. *Conservation Biology, 39*(1), 1–9. https://doi.org/10.1111/cobi.14328

Yazar, M., & York, A. (2023). Nature-based solutions through collective actions for spatial justice in urban green commons. *Environmental Science & Policy, 145*, 228–237. https://doi.org/10.1016/j.envsci.2023.04.016

Yazdi, A., & Karimian, A. (2021). Mapping Vaqf-Ābād Qanāt watercourse in the urban landscape of Yazd City in two distinct periods: Fourteenth and twentieth century. *Water History, 13*(1), 473–491. https://doi.org/10.1007/s12685-021-00290-8

Zarei, M., & Shahab, S. (2025). Nature-based solutions in urban green infrastructure: A systematic review of success factors and implementation challenges. *Land, 14*(4), 818. https://www.mdpi.com/2073-445X/14/4/818

Dialectical Relationship Between Human and Nature: Sustainable Placemaking in Hilly Cities with a Critical Perspective

Hooshmand Alizadeh, Milad Zandsalimi, and Milad Doostvandi

Abstract Rapid urbanization, driven by neoliberal ideologies, demands a balanced human–nature relationship to mitigate ecosystem degradation while integrating natural and built environments. Hilly urban regions, such as those in Sanandaj, Iran, exemplify this coexistence, embodying the dialectical interplay between human and natural spaces. This study proposes a theoretical framework for sustainable placemaking in hilly cities by critically re-examining relevant concepts and analyzing the socio-political forces that shape urban ecological systems and placemaking processes. Employing a multi-method approach—combining a theoretical literature review, field observations, stakeholder interviews, and GIS-based spatial analysis with four representative hills as case studies—this research utilizes MAXQDA for qualitative data analysis and ArcGIS Pro for geospatial mapping. The findings reveal two key insights: first, an analysis of land appropriation shows that, of Sanandaj's 60 hills, 32 are fully developed by the formal sector, 20 are claimed by governmental institutions, 7 by informal residents, and only 1 remains unappropriated. These are categorized into public green spaces, ecological zones, barren lands, and human-altered areas. Second, qualitative interviews with government officials, environmental activists, local residents, and investors indicate that formal power structures prioritize economic development, while communities value the hills' cultural, spiritual, and recreational roles. These dynamics highlight how unsustainable development and land speculation degrade ecosystems, exacerbate poverty, and reinforce spatial inequality. To address these challenges, this study advocates for a justice-oriented approach that engages local communities in participatory planning to foster sustainable placemaking. Such sustainability extends beyond preservation,

H. Alizadeh
Faculty of Art and Architecture, Department of Urban Planning and Design, University of Kurdistan, Sanandaj, Iran

Department of Geography and Regional Research, University of Vienna, Vienna, Austria

M. Zandsalimi (✉) · M. Doostvandi
Faculty of Art and Architecture, Department of Urban Planning and Design, University of Kurdistan, Sanandaj, Iran
e-mail: m.zandsalimi@uok.ac.ir

A. Cheshmehzangi et al. (eds.), *Designing Nature into Cities*, Urban Sustainability, https://doi.org/10.1007/978-981-95-9382-8_7

positioning hills as spaces for learning and communal well-being, achievable through intergenerational and intersectoral equity and resistance to appropriation by formal and informal powers. Ultimately, this research underscores the potential of equity-based strategies to redefine human–nature relationships and promote resilient, ecologically balanced urban futures.

Keywords Dialectical relationship · Human–nature interaction · Critical urban theory · Sustainable placemaking · Hilly cities · Sanandaj · Spatial justice

1 Introduction

Contemporary urbanization, driven by relentless expansion and neoliberal ideologies, stands as a defining feature of the twenty-first century, fundamentally reshaping the social, economic, and environmental foundations of human societies. Fueled by free markets and capital accumulation, this process transforms cities into arenas of global competition and ceaseless development, yet often strains and disrupts the human–nature relationship (Peck & Theodore, 2023). In 2022, approximately 57% of the global population—or 4.5 billion people—resided in urban areas, with projections indicating an increase to 68%, or 6.7 billion, by 2050 (UN DESA, 2022). The World Bank (2022) reports that between 2021 and 2022, 65% of infrastructure projects in hilly regions of South and Southeast Asia were designed to attract private investment, resulting in an 18% loss of natural vegetation cover. Hilly and mountainous regions, spanning about 24% of the Earth's land surface (approximately 35.5 million square kilometers), are home to over 1.1 billion people (FAO, 2023). Given that more than half are likely to reside in urban or peri-urban areas, an estimated 570–660 million urban dwellers inhabit montane cities (Huddleston et al., 2023). Capital-driven trends, accompanied by the degradation of hilly ecosystems, exacerbate pressures on natural resources and spatial inequalities, necessitating a rethinking of urban development models.

One of the most evident consequences of this rapid urbanization is the proliferation of informal settlements, particularly in hilly regions where land constraints and inequitable development policies are more pronounced (Davis, 2006; Roy, 2011). These settlements, often established on slopes and rugged terrain, result from widespread rural-to-urban migration, the absence of inclusive urban planning, and the dominance of capitalist logic in resource allocation (Bayat, 2013). According to UN-Habitat (2024), over 1.2 billion people worldwide live in such settlements, a significant portion of which are located in hilly regions, where the lack of basic infrastructure—such as potable water, sanitation systems, and access to healthcare—exacerbates residents' vulnerability (UN-Habitat, 2024). This situation not only places additional pressure on fragile ecosystems but also deepens spatial and social inequalities (Soja, 2010). Research indicates that vegetation loss and soil erosion due to informal construction are up to 35% higher than in flat areas (Gupta et al., 2023), highlighting the need to rethink urban development models and prioritize environmental justice (Agyeman et al., 2021). These challenges reflect a fundamental tension between humans and nature, embodying a dialectical dynamic.

The human–nature relationship, amidst relentless urban expansion, evolves into a dynamic dialectical struggle where humans shape nature, and nature, in turn, redefines patterns of human life, culture, and survival (Harvey, 2023). This relationship is not unidirectional but a reciprocal and evolving interaction in which both sides simultaneously act as creators and creations of each other. Understanding this interplay becomes especially crucial within hilly urban contexts, where natural constraints and human needs are deeply intertwined (Lefebvre, 2021). These areas, characterized by complex topography, fragile ecosystems, and unique geographical features, offer opportunities for creating distinctive and sustainable places—but they are also increasingly exposed to serious ecological and social threats due to pressures from urban development (Kaur & Garg, 2023). Urban hills serve as sites where human and natural activities converge, embodying the ceaseless interplay between humans and nature. Human survival on Earth requires controlled interventions to mitigate ecosystem degradation, yet this control does not entail preserving these landscapes in a pristine state within modern urban contexts—particularly in cities where capital decisively shapes spatial configurations. However, rapid urbanization often reduces direct engagement with natural landscapes, fostering an "extinction of experience" that leads to environmental amnesia, particularly among contemporary generations in modern metropolises (Kahn Jr. & Kellert, 2002).

The dialectical human–nature relationship in hilly cities is shaped by sociopolitical forces that govern resource allocation, knowledge production, and capital distribution (Harvey, 2018). Power, concentrated among formal institutions and informal actors such as land speculators, often commodifies nature, framing hills as exploitable resources (Peck & Theodore, 2023). In contrast, indigenous and local knowledge, rooted in communities' lived experiences, recognizes these spaces as cultural assets and as vital for survival (Escobar, 2020). This tension between conflicting representations underscores the need to rethink placemaking processes, shifting away from capital-centric domination toward inclusive stakeholder engagement (Healey, 2022). Sustainable placemaking transcends mere conservation, aiming to create spaces where humans see themselves as integral to ecosystems, achieving resilience and equity through collective collaboration (Agyeman, 2022). Sustainable placemaking and the preservation of natural environments extend beyond parks and highway margins, encompassing ecosystems defined by their full range of services, which enhance urban quality of life across all dimensions (Yu et al., 2001). Ultimately, the sustainability of ecosystems, through their protection from overexploitation and integration with human activities, is meaningful only when a balanced coexistence is achieved.

As a process that redefines human–nature interactions, sustainable placemaking emerges as both an environmental goal and a critical tool for challenging dominant development paradigms (Brenner, 2022). While prior studies offer valuable insights, they reveal significant gaps. For example, research by Harvey (2023), Swyngedouw (2022), and Heynen (2019) examines dialectical human–nature interactions in urbanization, yet these studies rarely address the unique challenges posed by hilly regions. Studies on sustainable placemaking (Healey, 2022; Agyeman et al., 2021; Folke et al., 2023) emphasize ecological and social balance, yet they largely focus on flat urban areas and overlook topographic constraints. Political economy analyses (Peck & Theodore, 2023) examine the influence of power and capital, yet they fail to comprehensively address hilly contexts. Studies of hilly regions (Kumar et al., 2022; D'Angelo et al., 2022) frequently address ecosystem sustainability, yet they lack a robust dialectical framework. Similarly, work on participation and indigenous knowledge (Berkes, 2018; Escobar, 2020; Sandercock, 2021) and spatial justice (Soja, 2010; Roy, 2011; Schlosberg, 2022) emphasizes community roles, yet seldom integrates political economy or focuses on hilly regions specifically.

These gaps, including the absence of an integrated, dialectical analysis; limited attention to hilly characteristics; and a lack of participatory planning solutions, underscore the need for a comprehensive, critical framework for sustainable placemaking in hilly settlements. Conventional approaches often treat human–nature interactions unilaterally, overlooking critiques of power and capital structures, thereby reproducing vulnerabilities (Gupta et al., 2023). A critical perspective grounded in political ecology and neoliberal urbanization theories reimagines sustainable placemaking as a liberatory act that can resist capital dominance and environmental degradation (Peck & Theodore, 2023). Unlike mainstream models that focus on technical resilience or aesthetics, this perspective examines how political-economic forces shape hilly spaces and seeks to redefine them through community participation and indigenous knowledge (Ramirez et al., 2021). Without such an approach, as recent research shows, development in hilly regions often leads to ecosystem destruction and greater spatial inequality (D'Angelo et al., 2022).

Sanandaj, the capital of Iran's Kurdistan Province and known as the "city of a thousand hills," is an example of these dynamics (Alizadeh, 2021). The city was founded atop Officers' Club Hill, located in a basin between the Abidar Mountains to the west and the Salavatabad Heights to the east. Sanandaj encompasses more than 60 hills within its legal boundaries (Alizadeh et al., 2019). Although the 2019 Sanandaj Hills Management Plan, as a high-level policy, establishes a public right and legal obligation to protect these hills, it requires updated studies, particularly those addressing development trends, to foster sustainable placemaking. This study examines existing literature gaps and proposes a critical framework for sustainable placemaking in hilly settlements, targeting not only environmental sustainability but also social justice and resistance to development hegemony. Its objective is to provide a theoretical and practical framework for sustainable placemaking in hilly cities, balancing human needs with natural contexts. Employing a mixed-methods approach—including theoretical literature reviews, GIS-based graphical analyses to monitor hill changes, participatory field observations with local communities, and interviews with key stakeholders—this study explores the multifaceted dialectical human–nature relationship in hilly settlements. The research is guided by two primary questions:

1. How can the dialectical relationship between humans and nature be leveraged to create sustainable places in hilly cities?
2. What socio-political factors influence the process of sustainable placemaking in the hills of Sanandaj?

2 Background Knowledge

2.1 *Political Economy of Space*

The political economy of space provides a critical analytical framework for understanding the interplay of power, capital, survival, and knowledge in the production and representation of urban environments (Peck & Theodore, 2023; Blomley, 2021). Power, concentrated within both formal institutions and informal actors, shapes how natural spaces—such as hills—are represented and utilized. Guided by neoliberal logic, these actors often frame nature as a resource for capital accumulation, transforming hills into sites for infrastructure development, mineral extraction, or residential construction (Gupta et al., 2023).

This dominance of power restricts local communities' access to resources and marginalizes their role as producers of knowledge. Consequently, representations of nature are often engineered to benefit specific interest groups, resulting in ecosystem degradation and heightened social inequality (Swyngedouw, 2022; Berkes, 2012). Knowledge functions as a dual variable: technical knowledge, produced by formal institutions and experts, frequently legitimizes development-oriented projects while downplaying their environmental and social consequences (Harvey, 2023). In contrast, indigenous knowledge, rooted in local communities' lived experiences, offers sustainable practices for managing resources in ways that align with the ecological characteristics of hilly areas (Berkes, 2012). Despite its significant ecological value, indigenous knowledge is frequently marginalized—a direct reflection of underlying power asymmetries (Kumar et al., 2022).

Capital, as another driving force, intensifies these tensions. Public and private investments frequently prioritize short-term economic gains, reducing nature to a commodity valued primarily for its financial profitability (Peck & Theodore, 2023). Human survival, however, depends on redefining this relationship—one that encompasses not only physical existence but also the preservation of cultural identity and the enhancement of resilience in the face of ongoing socio-ecological challenges (Folke et al., 2023). In hilly urban settlements, rethinking the political economy of space requires redistributing power from centralized institutions to local communities, thereby enabling their active participation in decision-making processes and leveraging indigenous knowledge for sustainable resource management (Smith, 2023; Brown et al., 2022). Capital must shift from accumulation-driven logics toward sustainability-oriented strategies, ensuring that investments concurrently support ecosystem preservation and socially responsible infrastructure development (Agyeman, 2022). Ultimately, human survival in these contexts hinges on integrating equitable power relations, hybrid knowledge systems that combine indigenous and technical expertise, and capital reinvestment that fosters a balanced coexistence between humans and nature (Escobar, 2020). Such an integrated approach can transform hilly urban settlements into sustainable spaces that serve as models for resilient and equitable human–nature coexistence.

2.2 Human–Nature Relationship

The relationship between humans and nature is inherently dialectical—a dynamic process in which both parties mutually influence and redefine each other (Harvey, 2023; Castree, 2021). Rooted in philosophical and geographical thought, this perspective views nature not as a static backdrop but as an active context continually shaped by human interaction (Lefebvre, 2021). In hilly urban areas, this dialectical relationship can be understood through four distinct theoretical approaches:

1. The Conservationist Approach: This approach prioritizes the preservation of nature as an inviolable realm, insulated from human intervention, in line with traditional conservation policies (Folke et al., 2023). While it effectively safeguards ecological integrity, it often marginalizes the livelihood needs of local communities, creating significant tensions in urban contexts (Agrawal & Redford, 2022). For instance, the establishment of protected areas in India's Western Ghats restricted local access to forest resources, thereby undermining indigenous livelihoods (Bawa et al., 2004). In hilly cities such as Sanandaj, this approach impedes necessary development by overlooking economic imperatives and social equity, rendering it impractical within dynamic urban settings (Adams & Hutton, 2007).
2. The Power–Capital-Oriented Approach: This approach treats nature as an obstacle to progress, framing hills as barriers to urban expansion that must often be excavated or developed (Swyngedouw, 2022). Guided by neoliberal ideologies, it commodifies natural resources—such as soil, water, and vegetation—for economic profit, severing them from their ecological functions (Peck & Theodore, 2023). For example, hill leveling for real estate projects in Bengaluru, India, has led to severe soil erosion and biodiversity loss (Nagendra et al., 2018). In hilly cities like Sanandaj, such practices prioritize capital accumulation at the expense of ecological balance and community welfare, resulting in significant ecosystem degradation and deepened social inequalities (Gupta et al., 2023; Moore, 2021).
3. The Survival-Oriented Approach: This approach conceptualizes a symbiotic relationship between humans and nature, wherein mutual survival depends on sustainable and reciprocal interactions (Escobar, 2020). It draws heavily on indigenous knowledge, emphasizing ecological equilibrium through the integration of human livelihoods with environmental preservation (Tsing, 2023). For example, traditional Andean agroforestry practices in Peru maintain soil fertility and biodiversity (Zimmerer, 2011). While this approach fosters ecologically balanced green spaces in hilly urban settings, it often lacks the scalability required to address complex urban systems, as in Sanandaj, and remains limited in its engagement with broader urban political and economic structures (Tsing, 2023).
4. The Holistic Balanced–Controlled Approach: This approach frames humans and nature as interdependent partners, essential for mutual survival and well-being (Folke et al., 2023). Grounded in principles of environmental justice and indigenous knowledge, it redefines hills as active collaborators in sustainable place-making and advocates for equitable resource management through meaningful

community participation (Escobar, 2020). In Ethiopia's Tigray region, for instance, community-led terracing initiatives successfully restored degraded hills, enhancing both ecological resilience and social equity (Reij & Garrity, 2016). Applied to hilly cities like Sanandaj, this approach seeks a balance between preservation and development, fostering the creation of eco-social spaces—such as accessible public green areas—that help reduce spatial inequalities (Agyeman et al., 2021; Schlosberg, 2022).

These four perspectives highlight the fundamental tensions inherent in the human–nature relationship, particularly within hilly urban areas. The conservationist approach protects ecosystems but neglects essential human needs. The power–capital-oriented approach accelerates ecological degradation and exacerbates inequality, as seen in cases like Bengaluru. The survival-oriented approach promotes ecological balance but lacks scalability within complex urban systems. In contrast, the holistic balanced–controlled approach—exemplified by initiatives in Tigray—integrates community participation with environmental justice, making it the most suitable and transformative model for sustainable placemaking in hilly cities such as Sanandaj. Accordingly, this study adopts the holistic balanced–controlled approach for its capacity to co-create equitable and sustainable spaces, challenge dominant power structures, and constructively redefine urban human–nature dynamics (Agyeman et al., 2021; Escobar, 2020).

2.3 *Concept of Sustainable Place*

Grounded in urban studies and political ecology, sustainable placemaking goes beyond simple nature conservation to create spaces that balance human needs, ecosystem health, and cultural values (Healey, 2022; Jones et al., 2020; Alizadeh et al., 2024). Developed in response to unsustainable development and environmental degradation, this concept emphasizes the creation of places that are environmentally resilient, socially just, and culturally meaningful (Agyeman, 2022). In hilly cities, sustainable placemaking is especially critical due to complex topographies and fragile ecosystems, which naturally limit uncontrolled development while offering unique opportunities to redefine human–nature relationships (Folke et al., 2023).

Hills, with their inherent natural landscapes, water resources, and biodiversity, hold significant potential for fostering multifunctional landscapes that meet human needs while maintaining ecological balance (Zhang et al., 2023; Chen et al., 2020). Sustainable placemaking in these contexts necessitates a shift away from centralized, profit-driven planning models toward genuinely participatory approaches that engage stakeholders, particularly local communities. Indigenous knowledge, rooted in residents' lived experiences, offers valuable methods for managing natural resources in ways that are adapted to the specific constraints and opportunities of hilly terrains (Berkes, 2012). Moreover, when properly managed, hills can enhance resilience to environmental challenges such as landslides, thereby transforming them into vital spaces that strengthen cultural and social identities (Smith, 2023).

This study argues that sustainable placemaking in hilly areas requires a fundamental shift in perspective: viewing hills not as obstacles or inert resources but as active partners in the development process. Such a perspective integrates environmental, social, and cultural values to foster spaces that harmonize human–nature interactions (Escobar, 2020). Within this framework, urban planners and designers play a pivotal role, acting not as top-down enforcers of power and capital-driven policies but as facilitators and co-producers who blend indigenous knowledge with scientific expertise (Pierce et al., 2022; Taylor et al., 2021). For example, in South Asian hilly regions, the integration of traditional techniques like terracing and rainwater management has been shown to reduce soil erosion by up to 30% while simultaneously preserving biodiversity (Kumar et al., 2022). By prioritizing collaboration with local stakeholders over rigid centralized structures, planners can significantly enhance urban resilience to climate change and natural hazards (Walker et al., 2019).

Ultimately, sustainable places function as critical tools to challenge and transform unsustainable development paradigms. Through inclusive community participation, they help reduce socio-ecological vulnerabilities and promote equitable outcomes (Sandercock, 2021; Reid et al., 2019). Figure 1 illustrates the dynamic interplay between humans and nature, highlighting how four key structural forces—capital, power, knowledge, and survival—interact to shape hillside urban spaces. The central zone represents a state of balanced power, achievable when competing stakeholder interests reach an equilibrium. The figure maps four distinct approaches to spatial representation along a conceptual continuum from anthropocentric to ecocentric worldviews, each guiding urban form differently. A clockwise circular arrow depicts the continuous feedback loop between societal actions and the natural environment, emphasizing the co-evolutionary and interdependent nature of these interactions over time.

3 Research Methodology

This study employs a mixed-methods approach, integrating both quantitative and qualitative data, to examine the dialectical relationship between humans and nature in Sanandaj—a city characterized by its distinctive hilly topography. The research seeks to construct integrated theoretical and operational frameworks for sustainable placemaking within urban environments influenced by complex interactions of power, socio-economic structures, and environmental dynamics. An analytical and descriptive framework, rooted in dialectical materialism, guides the investigation. This approach facilitates the identification of contradictions, interactions, and mutual transformations between human and natural systems (Heynen, 2019; Castree, 2021). The chosen methodology is aligned with the study's dual theoretical and practical objectives, enabling a nuanced understanding of multi-stakeholder interactions in hill appropriation and the processes of placemaking within a contested urban terrain.

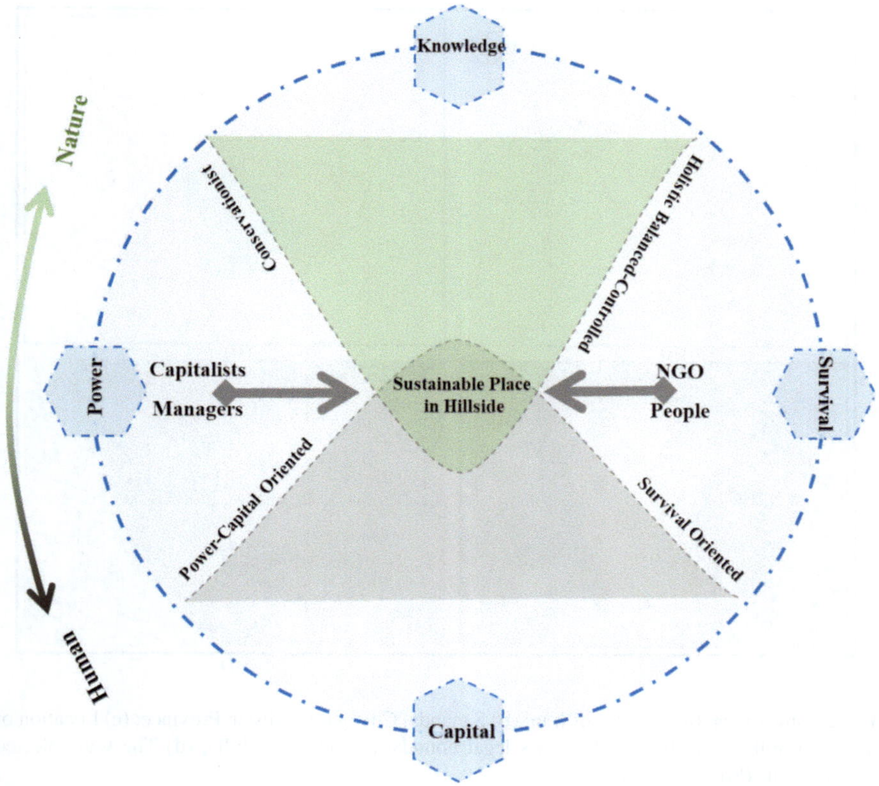

Fig. 1 Conceptual framework illustrating the dialectical relationship between humans and nature in hillside contexts, highlighting the structural forces of power, capital, knowledge, and survival (Source: Authors)

3.1 Case Study

Sanandaj, the capital of Iran's Kurdistan Province, serves as a significant urban case study renowned as the "City of a Thousand Hills" (Alizadeh, 2021). Situated within a topographic basin bounded by the Abidar and Kuchak-Rahsh mountain ranges (Bahrami et al., 2014), the city's distinct hilly terrain is a defining feature. The legal boundaries of this study, which encompass 60 hills, are formally defined by the 2019 Sanandaj Hills Management Plan, with 27 of these hills designated for detailed investigation (Bahrami et al., 2019). These administratively defined boundaries form the foundational spatial unit of this research and are supported by publicly available planning documentation.

The initial phase of the investigation involved a comprehensive survey of all 60 hills. Based on their dominant land use and observable human–nature relationships, the hills were categorized into four distinct typologies: Public green spaces,

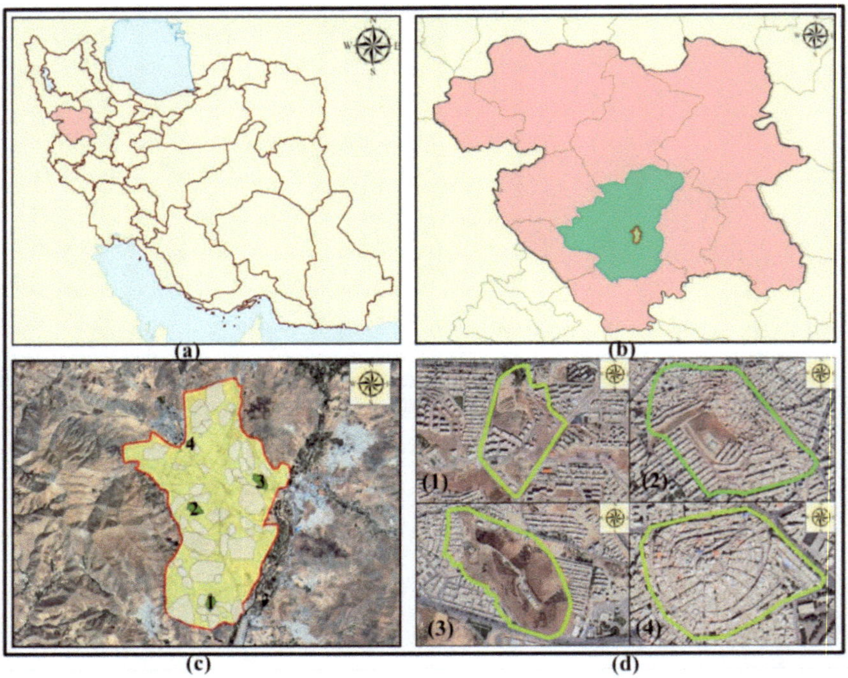

Fig. 2 Study area: (**a**) Country of Iran; (**b**) Sanandaj City and Kurdistan Province; (**c**) Location of the study hills in relation to the city's legal boundaries and its 60 hills; (**d**) The four selected research hills (Source: Authors)

Ecological conservation areas, Barren or undeveloped lands, and Built-up or human-constructed areas.

From this comprehensive set, four representative hills—Mir-e-Zurah, Keshavarz, Faridouni, and Sheikh Mohammad Sadegh—were selected for in-depth, multi-method analysis (see Fig. 2). These specific sites were chosen to reflect the full spectrum of human–nature interactions identified across the city, ranging from preserved public green spaces to intensively built-up areas. Each hill exemplifies significant spatial, social, and environmental transformations shaping Sanandaj's urban landscape. The selection criteria prioritized typological diversity and a range of socio-environmental dynamics, ensuring that these case studies are ideal for examining the dialectical processes inherent in sustainable placemaking.

3.2 Data Collection and Analysis

Data collection relied on two primary methods: GIS-based spatial analysis and semi-structured interviews. These methods addressed the research questions on leveraging the dialectical human–nature relationship for sustainable placemaking and identifying socio-political factors in Sanandaj's hills (Creswell & Plano Clark, 2018). For GIS analysis, three data sources were used: (1) the Sanandaj detailed urban plan (approved 2013), including land use maps and the hill organization plan for the 60 designated hills, sourced from the Roads and Urban Development Department of Kurdistan; (2) Google Earth Pro imagery (2020, version 7.3.6) to map land use zones (built-up, barren, and green spaces); and (3) field observations from March to May 2025 across all 60 hills, with intensive surveys on four case study hills—Mir-e-Zurah, Fereydouni, Sheikh Mohammad Sadegh, and Keshavarz. Each hill was visited at least twice, producing at least 20 photographs and detailed notes for the case study hills. Field coordinates were cross-checked with Google Earth imagery and urban plan maps. Land use zones were manually delineated using the ArcGIS Pro 3 Digitizing tool via overlay analysis (Longley et al., 2015).

For interviews, 45 participants were purposively selected based on their experience or knowledge of the hills and willingness to participate. The group included 8 government officials (from the Roads and Urban Development Department of Kurdistan, Sanandaj Municipality, Natural Resources, and Environment departments), 5 environmental activists (NGO representatives and researchers), 20 hill residents (diverse in age, gender, and location), and 12 real estate investors (developers and brokers). Theoretical saturation was reached when thematic analysis revealed no new concepts across stakeholder groups (Saunders et al., 2018). Interviews, lasting 30–60 min, used common questions for all groups, with tailored questions for each stakeholder type. They were audio-recorded with consent, transcribed verbatim, and analyzed using thematic analysis to identify key socio-political and dialectical themes (Braun & Clarke, 2006).

GIS-based graphic analysis compiled hill study maps, integrating results with field observations, existing land use patterns, and meta-studies. Observed land use patterns were categorized into three classes: barren lands, green spaces, and built-up areas. MAXQDA 2024 software analyzed qualitative data, employing tools like Code Frequency, Word Cloud, Matrix Browser, Co-occurrence Analysis, and Concept Map. These tools identified core themes, analyzed semantic relationships, and visualized connections between categories, aligning with the findings section.

4 Results

4.1 Classification of Sanandaj Hills Based on the Dialectical Relationship Between Human and Nature

The occupation patterns of Sanandaj's hills reflect a struggle for control among three key actors: (1) governmental and military institutions, (2) formal developments by middle- and upper-class groups through official planning and capital accumulation, and (3) informal settlements by low-income populations without legal documentation. Of the 60 hills within the urban boundaries, 32 (53%) are dominated by construction, driven mainly by formal development programs, leading to significant ecological degradation. Among the remaining 28 hills, where at least the peak, mid-slope, or base remains unappropriated, 20 are occupied by formal sectors, 7 by informal sectors (now within legal boundaries), and one (Qout) remains unclaimed. Of the 20 formally occupied hills, 13 peaks are controlled by powerful institutions or governmental bodies, while other constructions at the bases include both planned and self-initiated structures, contributing to hill degradation.

Informal settlements, formed by rural migrants from surrounding villages, exhibit four traits: lack of ownership documents, low-quality materials, low-income residents, and narrow access roads. Due to limited financial and technical resources, earth excavation in these areas is lower than in formal developments, which follow urban plans and allocate land officially. Land prices in informal zones are lower than elsewhere in Sanandaj. Transporting materials to hilltops is challenging, and narrow streets hinder vehicle access, causing price variations between hill bases, mid-slopes, and tops—especially where buildings exist.

Based on the dialectical human–nature relationship, Sanandaj's hill settlements are classified into four categories: public green spaces, ecological areas, barren lands, and human-occupied areas (Fig. 3a). These are detailed below:

Fig. 3 (**a**) Classification of Sanandaj city hills into four main categories, illustrated with representative images; (**b**) Accumulation of buildings in the informal settlements of Abbas Abad and Farajeh; (**c**) The appropriation of Be'sat Hill's peak; (**d**) Destruction caused by development proposals on Tepe Degairan; (**e**) The natural cover of Nishtman Hill (Source: Map and photos by the Authors)

1. Hills with Public Green Spaces

These hills feature significant vegetation (e.g., native shrubs, grasses) and serve as urban public spaces. Residents use them freely for leisure, picnics, or social gatherings. Four hills—Melat, Didgah, Mir-e-Zurah, and Jabal-al-Rahma—offer amenities like benches and pathways, functioning as viewpoints or recreational areas. However, Jebel Rahmeh's green spaces suffer from poor social performance due to industrial pollution from nearby dense, informal Abbasabad settlements (Fig. 3b).

2. Hills with Ecological Green Spaces

These hills are characterized by forest vegetation and wildlife and hold strong ecological value. Surrounded by urban roads, the hills function as isolated green islands, disconnected from wider public activities. Five hills—Tous Nowdhar, Keshavarz, Rousi, Pir Gharib, and Nishtman—fall into this category (Fig. 3e). Although designated areas for public use exist on the peaks of Pir Gharib and Tous Nowdhar, their classification as ecological spaces is maintained due to the dominance of vegetation and ecological function. Terraced residential constructions have been established at the mid-slopes and bases of Pir Gharib by migrants from villages such as Galin in the Uramanat region. These constructions lack coherent architectural patterns and reflect informal, self-built processes.

3. Barren and Abandoned Hills

These hills are mostly undeveloped and neglected land left behind by urban expansion. Portions of their bases, slopes, and peaks remain unused. The failure of proposed urban development plans, particularly those intended for service-oriented land uses, is largely due to the neglect of topographic constraints. Governmental and military institutions have also contributed to the abandonment of these areas through ownership and power over construction activities (Fig. 3c). Fourteen hills, such as Sanjili, Dan, Fereydouni, Takieh, and Chaman, belong to this category.

4. Hills with Accumulated Human-Made Elements

In these cases, hills are buried beneath dense layers of construction. Extensive excavation and human intervention have eliminated natural elements, such as springs, vegetation, and wildlife (Fig. 3d). The economic priorities of construction and the shelter demands of low-income communities have driven these transformations. Some hills, such as Pir Salam, Sheikhān, Sheikh Mohammad Baqer, and Sheikh Mohammad Sadeq, were historically converted into cemeteries. Today, their peaks and mid-slopes are often used by criminals and drug users, transforming potential public spaces into sites of insecurity. Sheikh Mohammad Sadeq Hill has seen some improvement, though encroachment continues. The cultural and historical significance of the tombs of Sheikh Mohammad Baqer and Pir Salam, as well as the historic graves on Sheikhān Hill, has protected these sites from large-scale construction thus far. However, recent non-residential developments at their bases have increased insecurity, especially at night.

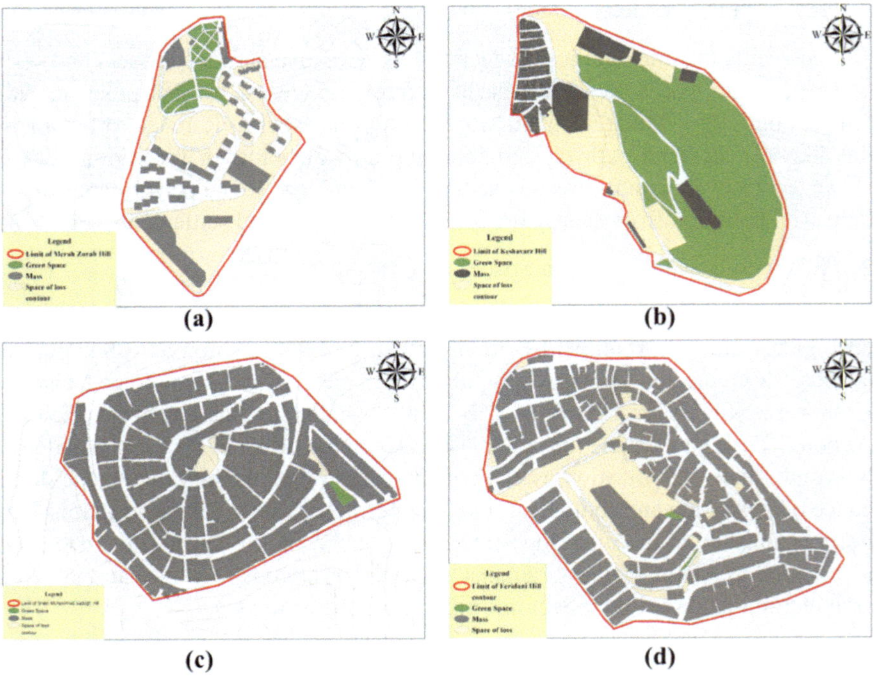

Fig. 4 Classification of study hills; (**a**) Mir-e-Zurah Hill; (**b**) Keshavarz Hill; (**c**) Sheikh Mohammad Sadeq Hill; (**d**) Fereydouni Hill. (Source: Authors)

Of the proposed classification, the four hills of Mir-e-Zurah, Fereydouni, Sheikh Mohammad Sadeq, and Keshavarz were selected as case studies. The hills were analyzed in terms of barren land, human-made developments, and remaining green spaces. The analysis also considered the political, social, and economic forces that shaped their transformation and spatial representation. Except for Mir-e-Zurah, the other three hills have been significantly shaped by the housing demands of low-income communities (Fig. 4). Keshavarz has the highest level of green space, with 60% forest cover, while Sheikh Mohammad Sadeq has the lowest, at only 5% (approximately 2000 m^2). Over 50% of Fereydouni and Sheikh Mohammad Sadeq are occupied by human-made structures (25,000 and 30,000 m^2, respectively). This extensive construction has produced high residential density—approximately 200 persons per hectare—driven by fine-grained urban plots averaging 100 m^2. In contrast, Fereydouni retains a notable proportion of barren land (40%), largely due to land hoarding by powerful actors, including government and military institutions (Table 1).

Except for Mir-e-Zurah, the spatial development of Fereydouni, Sheikh Mohammad Sadegh, and Keshavarz has been shaped by dominant elements at their peaks. A military center atop Fereydouni, an incomplete tourism complex on Keshavarz, and historical graves on Sheikh Mohammad Sadegh have influenced their spatial layout (Fig. 5). On Sheikh Mohammad Sadegh, informal settlements

Table 1 Examination and comparison of the spatial and socio-economic characteristics of the four hills of Mir-e-Zurah, Fereydouni, Sheikh Mohammad Sadegh, and Keshavarz (Source: Authors)

Hills	Hill area (m²)	Category area (m²)		Percentage of area of hill	Socio-economic context
Mir-e-Zurah	207,450	Green Space	15,520.17	7.48	Apartment complexes with a design and geometry whose residents are from middle to upper income groups
		Mass	40,730.51	19.63	
		Barren lands	86,354.22	41.63	
		Sum	142,604.90	68.74	
Keshavarz	253,245	Green Space	131,522.7	51.93	Spontaneous settlements of Abbas Abad and Farajeh with relatively low-income levels and industrial, educational, military and facility facilities
		Mass	34,259.52	13.53	
		Barren lands	59,547.38	23.51	
		Sum	225,329.6	88.98	
Freidouni	250,643	Green Space	600.58	0.24	A combination of the low-income, spontaneous Ghafoor settlement on the western front and the checkered Balgo neighborhood of Saadi Town on the eastern front of the hill
		Mass	13,6801.1	54.58	
		Barren lands	32,863.09	13.11	
		Sum	170,264.8	67.93	
Sheikh Muhammad Sadiq	86,674	Green Space	426.4	0.49	Dense and spontaneous settlement of migrants in villages along the Sanandaj-Divandere route
		Mass	57,809.75	66.77	
		Barren lands	1138.04	1.31	
		Sum	59,374.19	68.50	

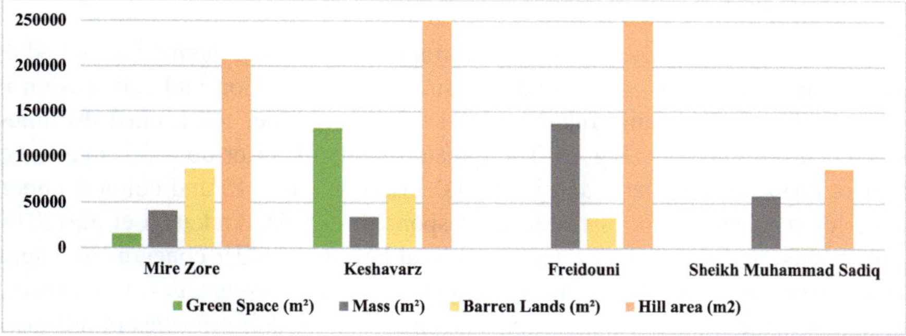

Fig. 5 Area values of green spaces, mounds, and wastelands compared to the total area of the hill in square meters (Source: Authors)

have led to constructions that often overlook historical graves, though some cemetery remnants remain intact. Mir-e-Zurah, once known for springs and green spaces per long-time residents, faced major degradation in the 1970s due to urban expansion for Baharan town. One high-flow spring still survives. The expanding informal settlements of Farajeh at Keshavarz's base, alongside the adjacent industrial city, threaten its vegetation and wildlife. Established in 1991 under the Farafza plan, this industrial area persists despite urban growth. Insecurity from Farajeh and Abbasabad settlements near Keshavarz prompted a police station (8200 m^2) on its western side. Proposals for educational facilities and a fuel station on the north, plus industrial and security uses on the west, show limited focus by planners on the hill's ecological value. Keshavarz's location at a main entry from central Iran, with forest vegetation near Tus Nozar and Jabal-al-Rahma, highlights its importance.

4.2 Perspectives of Stakeholders

This section presents the results of a qualitative analysis based on 45 semi-structured interviews with key stakeholders involved in Sanandaj's hills. These stakeholders included government officials, members of environmental associations, local residents, and investors. MAXQDA software and an inductive content analysis approach were used to analyze the interviews, with the objective of identifying the main themes related to the study's core variables. Several analytical tools were employed to support this process, including Code Frequency, Word Cloud, Matrix Browser, Co-occurrence Analysis, and Concept Mapping. These tools provided complementary insights into stakeholder perspectives and revealed patterns of meaning across different groups. The findings generated by these tools were systematically aligned with the initial results, offering a comprehensive understanding of the spatial and social dynamics shaping Sanandaj's hill settlements.

- *Code Frequency*

Code Frequency analysis showed "destruction" as the top theme, mentioned 45 times. This aligns with earlier findings, such as deforestation and soil erosion in Sanandaj hills. "Economic Interests" and "Cultural Value," each cited 40 times, were equally significant (Fig. 6). These match economic priorities of formal stakeholders (90% of managers' and investors' responses, Fig. 8) and cultural importance for residents (85% of residents' responses, Fig. 8). Anderson et al. (2015) support this, noting frequency analyses reveal key stakeholder concerns in natural resource management. By using numerical values and percentages, this analysis underscores the need to balance environmental and cultural priorities in hill management and planning.

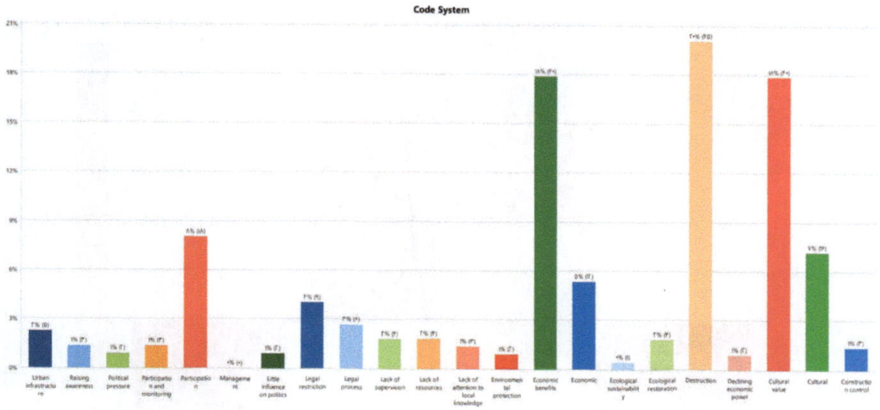

Fig. 6 Frequency of key codes from stakeholder interviews (Source: Authors)

Fig. 7 Word cloud of key themes from stakeholder interviews (Source: Authors)

- *Word Cloud*

The word cloud is a widely used tool in qualitative data analysis that helps to identify linguistic patterns, recurring concepts, and key themes within textual datasets. In this study, the word "degradation" appeared most prominently, with a frequency of 50 occurrences, which corresponds with the abstract's emphasis on environmental challenges such as deforestation and soil erosion. The words "identity" (35 times) and "culture" (25 times) were also frequently highlighted, reflecting, respectively, the economic pressures and the spiritual values associated with the hills for local residents. The visualization, which scales word size according to frequency, underscores the need to prioritize both environmental conservation and cultural values in sustainable placemaking (Fig. 7). The most significant keywords included identity, cultural value, degradation, lack of participation, and economic interests.

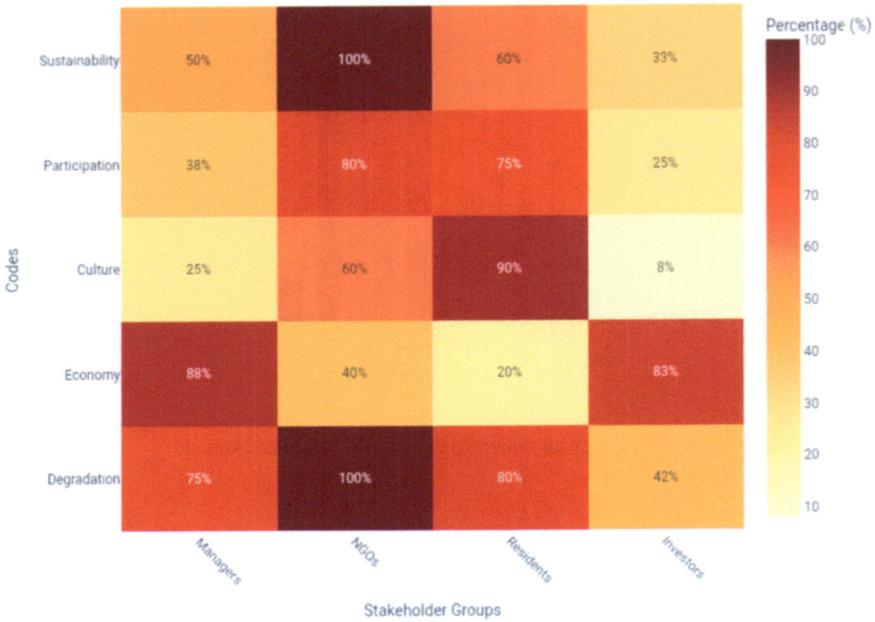

Fig. 8 Stakeholder perspectives related to selected keywords (Source: Authors)

- *Matrix Browser*

Coding trends, interconnections among data, and semantic patterns were examined across the collected documents. The results indicate that managers placed greater emphasis on development, whereas residents prioritized the value and identity of the hills. Environmental advocates highlighted preservation and ecological restoration, while entrepreneurs focused on the economic benefits of Sanandaj's hills. This matrix illustrates that the gap between official perspectives (which tend to view resources as exploitable) and local perspectives (which stress cultural value) obstructs sustainability, thereby reinforcing the need for justice-oriented approaches (Fig. 8). These findings are consistent with Folke et al. (2005), who demonstrated that conflicts between economic and ecological objectives in urban natural resource management undermine resilience. The matrix further underscores the necessity of integrated planning strategies to reduce inequalities in decision-making.

Additionally, the Code Relations Browser was used to analyze semantic connections and identify common patterns and code co-occurrence in the textual data. This analysis illustrated key relationships among different codes. For example, the code "degradation" was strongly related to "development," "environment," and "economic benefits." Moreover, a clear connection was found between "participation" and "cultural values" (Fig. 9). The strongest relationship was identified among "participation," "policy," and "cultural value," underscoring the vital role of community involvement and cultural considerations in shaping effective urban policies.

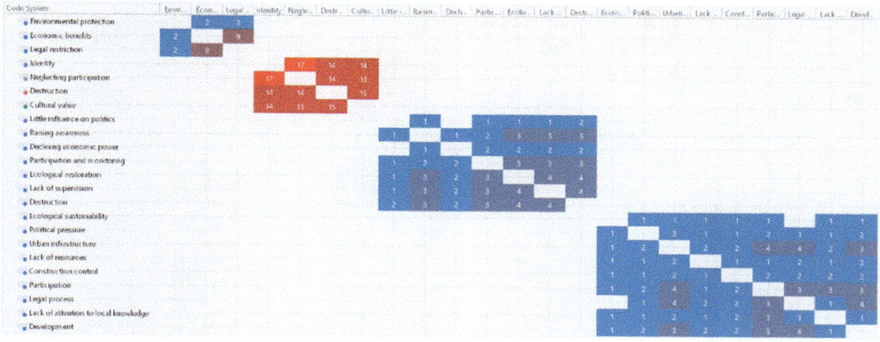

Fig. 9 Code Relations Matrix (Source: Authors)

- *Co-occurrence Analysis*

The co-occurrence analysis measures the intensity of code overlap using numerical scales and highlights the intersection of ecological, economic, and social dimensions in Sanandaj's hills. The strongest relationship was identified between "ecological degradation" and "economic interests," with a co-occurrence frequency of 71% (32 respondents). This finding aligns with preliminary results indicating that profit-driven construction accelerates deforestation and soil erosion. Another significant relationship emerged between "low participation" and "injustice" (60%, 27 respondents). This emphasizes the exclusion of local communities from decision-making processes and underscores the urgent need for participatory planning (Fig. 10). These results align with those of Elmqvist et al. (2013), who demonstrated that economic pressures and power inequalities pose significant challenges to the management of sustainable urban ecosystems. Together, these patterns reveal that addressing environmental degradation and social injustice is essential for achieving resilience in hilly urban settlements.

- *Concept Map*

The concept map illustrated the hierarchical structure of the study's themes in a tree-like format, identifying four main categories related to Sanandaj's hills: managerial, economic, environmental, and cultural. Within the managerial category, codes such as legal processes, construction control, legal constraints, policies, oversight, and economic pressures were identified. The economic category encompassed themes including degradation, urban infrastructure, resource scarcity, and development. The environmental category highlighted restoration and environmental sustainability, as emphasized by environmental activists and consistent with the preliminary findings regarding the destructive impacts of profit-driven activities on Sanandaj's hills. In the cultural category, the most frequent codes were identity, participation, awareness, and local knowledge, underscoring the importance of community-based solutions (Fig. 11). This structure aligns with Rees et al. (2014), who demonstrated that concept maps effectively represent complex interconnections between environmental and social factors. By visualizing themes as connected

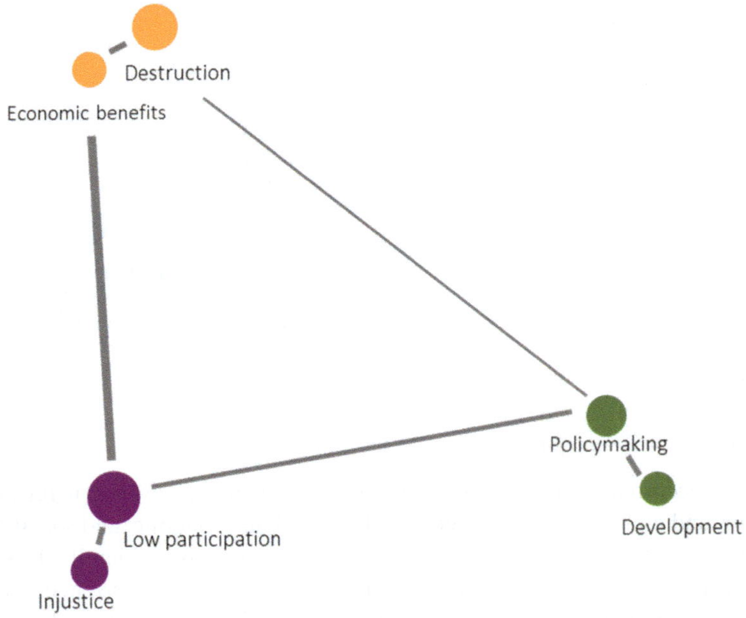

Fig. 10 Co-occurrence analysis of stakeholder responses (Source: Authors)

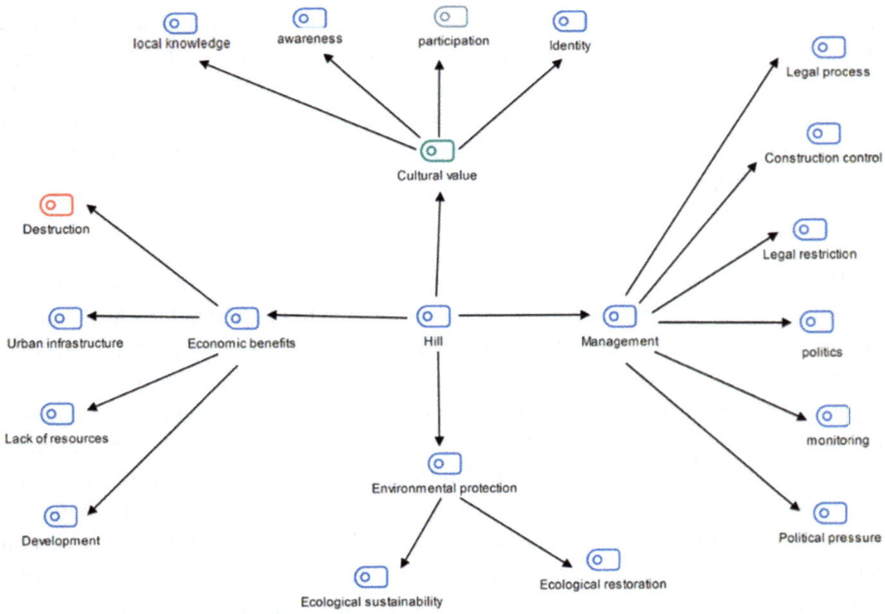

Fig. 11 Conceptual model aligned with extracted categories and their themes (Source: Authors)

nodes, the concept map underscores the necessity of integrating ecological preservation with community participation in sustainable urban planning.

5 Discussion and Implications

By synthesizing spatial dynamics with theoretical frameworks and stakeholder perspectives, this discussion draws on urban political ecology and environmental justice scholarship to propose transformative strategies. The analysis highlights how power structures, ecological vulnerabilities, and community resilience interweave to shape placemaking processes, advocating for equity-driven interventions that challenge exploitative paradigms in topographically complex urban settings.

The classification of Sanandaj's hills into public green spaces, ecological green spaces, barren lands, and human-made constructed areas illuminates the dialectical relationship between humans and nature, central to sustainable placemaking in hilly urban contexts. The prevalence of human-made constructed hills, driven by both formal developments and informal settlements, reflects a power-capital-oriented approach, where institutional control and economic priorities accelerate ecological degradation (Harvey, 2023). This pattern mirrors broader neoliberal urban expansion, where profit-driven development erodes natural landscapes (Peck & Theodore, 2023). By contrast, ecological green hills, such as Keshavarz, embody a conservation-oriented approach, preserving biodiversity but limiting public access, similar to protected ecosystems in Tigray (Reij & Garrity, 2016). Public green hills, like Mir-e-Zurah, align with inclusive approaches, fostering community engagement—though challenges such as industrial pollution in Jabal-al-Rahma constrain their potential (Agyeman et al., 2021). Informal settlements, such as Pir Gharib's terraced constructions, represent a survival-oriented approach, where marginalized communities adapt local knowledge for resilience despite exclusion (Escobar, 2020).

These findings suggest that sustainable placemaking can harness this dialectic by integrating ecological restoration with community-driven design, transforming barren hills into socio-ecological corridors. Similar outcomes have been observed in participatory hill restoration projects (Switzer et al., 2019). For instance, reforestation initiatives inspired by Inca practices could restore ecological integrity while fostering public access, potentially reducing soil erosion by up to 40% in degraded areas (Berkes, 2018; Zimmerer, 2011). This dialectic also opens opportunities for hybrid approaches that combine conservation with inclusion. For example, incorporating public access into ecological green hills could reduce urban heat islands by 2–3 °C while creating biodiversity corridors that support 20–30% more species diversity, as shown in comparable hilly urban restorations (Switzer et al., 2019). Globally, equitable placemaking models have increased community resilience by 25–35% in topographically challenging environments, underscoring the importance of policies prioritizing long-term ecological health over short-term economic gains (Folke et al., 2023; Alizadeh et al., 2024).

The qualitative analysis of stakeholder perspectives further reveals tensions between economic interests and ecological/cultural values, highlighting socio-political barriers to sustainable placemaking. Managerial narratives typically frame hills as exploitable resources, while residents perceive them as cultural anchors—echoing critiques of neoliberal governance, where institutional control prioritizes profit and security over ecological and cultural concerns, deepening spatial injustice (Brenner & Schmid, 2015; Kaur & Garg, 2023). In Sanandaj, these dynamics are intensified by military monopolies and economic pressures, which constrain community participation. For example, the co-occurrence of "degradation" and "economic interests" in stakeholder responses suggests that, without inclusive governance, urban policies could exacerbate environmental decline by 30–50% in vulnerable hilly areas (Gupta & Rao, 2023; UN-Habitat, 2024). To address these challenges, participatory models could empower residents to integrate cultural values into planning. Community-driven restoration projects have reduced injustice and enhanced social cohesion by 20% (Kumar et al., 2022; Healey, 2022). Similarly, participatory GIS can map ecological and cultural assets, reducing urban inequality by up to 25% (Ramirez et al., 2021). Such strategies challenge hegemonic structures and align with adaptive governance models that emphasize social-ecological resilience, potentially mitigating degradation by 30–40% in vulnerable hilly areas (Folke et al., 2023; UN-Habitat, 2024).

The concept map, which organizes themes into managerial, economic, environmental, and cultural categories, integrates spatial and socio-political dynamics to guide sustainable placemaking (Rees et al., 2014). The managerial category, reflecting legal constraints and economic pressures, highlights institutional barriers that reinforce power imbalances—oversight failures alone exacerbate degradation by 25% (Kaur & Garg, 2023). The environmental category emphasizes restoration, aligning with global initiatives that have increased urban green cover by 15–20% (FAO, 2023). The economic category reflects resource scarcity, underscoring the need for balanced strategies, as seen in inclusive models that reduced spatial disparities by 20% (Escobar, 2020). Finally, the cultural category—emphasizing identity, participation, and local knowledge—illustrates how community-based solutions can transform exclusion into resilience, enhancing cultural sustainability by up to 30% (Zimmerer, 2011). Based on the classifications in the first and second sections, which examined four main hill categories and extracted four managerial, cultural, environmental, and economic categories, respectively, the following proposals are offered for sustainable placemaking in Sanandaj's hills (Table 2).

Table 2 Proposals for sustainable placemaking in Sanandaj's hills across hill classification categories (Source: Authors)

Hill classification	Categories			
	Economic	Environmental	Cultural	Management
with public green spaces	Creating a participatory management system with local institutions	"Hold ceremonies, festivals, and cultural events to increase sense of belonging"	Pay attention to ecological functions alongside urban functions	Create indirect job opportunities through sustainable tourism and cultural services.
	Developing special criteria for public and service uses	"Use local knowledge and culture in improving and repairing spaces"	"Revive and develop vegetation with native species resistant to regional climate"	Branding the city's hills in line with the natural-economic context
with ecological green spaces	Attracting long-term investment in the field of natural resource conservation.	Strengthen the spiritual and cultural connection of residents with nature through educational and recreational programs	Launch a nature school	Increase monitoring and control of unauthorized construction around these hills.
	Designating protected areas and legal restrictions to prevent human interference.	Supporting local community participation in the protection of these areas.	Supporting sustainable agricultural and horticultural activities around these areas.	Paying ecological subsidies to local communities
Barren hills	Identifying and releasing wasteland owned by the government and military for public use.	Documenting oral history and indigenous subcultures	Reviving and developing vegetation in these areas with the aim of increasing urban respiration and reducing pollution.	"Create job opportunities through service, cultural, and tourism spaces"
	Modifying urban development plans according to the actual topography of the area.	Creating collective and cultural spaces in these areas to increase social and cultural interactions.	Creating green spaces with ecological functions alongside urban functions.	Attracting investment to improve and revitalize these spaces without ecological destruction.

(continued)

Table 2 (continued)

Hill classification	Categories			
	Economic	Environmental	Cultural	Management
with accumulation of anthropogenic elements	Improving the residential pattern in these areas through re-planning and implementing urban regeneration projects.	Creating community and cultural spaces in these areas to reduce social inequalities.	"Reduce environmental damage through pollution control, soil erosion, and waste management"	Supporting local and micro-economies in these areas through facilities and job creation.
	Increasing supervision and control over development and social activities in these areas.	Strengthening the cultural and spiritual connection of residents with the place through the preservation of existing historical and cultural monuments.	Creating ecological islands between buildings	Creating necessary infrastructure for sustainable economic development

6 Concluding Remarks

Sanandaj's hills, as stark symbols of the human–nature confrontation, are battlegrounds for an unequal struggle between urban development and environmental preservation. Findings indicate that power structures—formal institutions with unsustainable development plans and land speculators commodifying hills—not only destroy natural ecosystems but also perpetuate cycles of poverty and inequality through informal settlements. For instance, the degradation of 95% of hills in just a few decades testifies to the dominance of neoliberal development paradigms in urban spaces. This trend is accompanied by spatial injustice, where marginalized communities are both victims of top-down policies and, due to the need for shelter, unwitting contributors to degradation. At a broader scale, nature is the primary victim of this tragedy, and informal settlements, despite causing less destruction than capital and power holders, resort to hill slopes for survival, sometimes damaging vital organisms such as plant and animal species and springs. Yet, in their social and cultural constructs, they view hill preservation as self-preservation. The analysis of interviews confirms this paradox: urban managers view hills instrumentally as raw resources, while local residents see them as part of their collective identity.

This contradiction highlights the necessity of participatory and justice-oriented planning, where decisions are based on environmental, social, and cultural priorities. To break this vicious cycle, "dual strategies" must be employed: on one hand, redefining the human–nature relationship in urban planning discourse based on environmental ethics, and on the other, designing legal–social mechanisms to curb the power of capital and dominant institutions. The experience of Sanandaj's hills proves that passive preservation is insufficient; rather, a "critical restoration" of spaces must simultaneously address ecological restoration (e.g., reforestation) and social justice (e.g., organizing informal settlements with resident participation). The key lesson of this study is that sustainability in hilly urban areas requires moving beyond one-dimensional approaches: hills cannot be entirely degraded for development, nor can they be turned into natural museums without considering local community needs. Sanandaj can set a model for similar cities in Iran and globally by transforming its remaining hills into "ecological–social corridors" (combining green spaces, public uses, and sustainable housing).

In response to the first research question, the dialectical relationship for placemaking showed that of Sanandaj's 60 hills, 32 are under construction, 20 have been appropriated by formal entities (government and military), 7 have been informally seized, and only Qout Hill remains untouched. Hills were classified into four categories: public green (e.g., Mellat), ecological green (e.g., Tous Nozar), barren (e.g., Sanjili), and human-constructed (e.g., Feridouni). In the sample hills, Feridouni and Sheikh Mohammad Sadegh have dedicated over 50% of their area to constructions, Keshawarz has preserved the most green space, and Mireh Zoreh, despite 1970s degradation, retains an active spring. These data materially reflect the dialectical relationship: public and ecological green hills have established a relative balance between humans and nature, while barren and human-constructed hills have been degraded under human interventions. Interviews highlighted the potential of traditional knowledge (e.g., terraced settlement in Pir Gharib) and local participation for restoring barren and ecological hills. These findings align with Switzer et al. (2019), who examined Lisbon's hill management through local participation and ecological restoration. Sustainable placemaking in Sanandaj can be achieved through reforestation (e.g., tree planting), creating ecological parks, and integrating indigenous patterns into urban design, similar to Cusco's approach using Inca knowledge for hill management (Berkes, 2018). In response to the second research question, sociopolitical dynamics profoundly shape sustainable placemaking in Sanandaj's hills, revealing a dialectical tension between hegemonic power and community agency. Stakeholder interviews reveal that socio-political factors, including military monopolies, economic pressures, and low community participation, exacerbate hill degradation in Sanandaj, undermining sustainable placemaking. Neoliberal governance, critiqued by Brenner and Schmid (2015), drives institutional control and land commodification, prioritizing profit and security over ecological and cultural values, thus deepening spatial injustice. Marginalized communities, like those in Farajeh, navigate this exclusion through informal settlements, simultaneously disrupting ecosystems and asserting cultural ties to the hills. This paradox—where top-down policies erode sustainability while grassroots resilience offers hope—demands justice-oriented planning. Participatory frameworks, inspired by Bogotá's local councils (Escobar, 2020), can empower residents, integrating traditional knowledge with environmental justice principles (Angelovski et al., 2016) to reimagine hills as equitable, sustainable spaces.

This study faced limitations in data collection due to restricted access to some military-controlled hills, potentially underrepresenting their ecological and social dynamics. Additionally, the qualitative nature of the stakeholder interviews may limit generalizability, as perspectives may vary across other hilly cities. It is recommended for future research to examine Sanandaj's peripheral urban zones, where spontaneous informal settlements predominate, to advance comprehensive sustainable placemaking. Employing remote sensing technologies, such as satellite imagery and Google Earth Engine, alongside participatory GIS (PPGIS), could significantly enhance methodological precision and data robustness, fostering informed strategies for equitable urban planning and ecological restoration in hilly urban contexts. Comparative analyses with other hilly cities in Iran or globally could further refine strategies for sustainable placemaking. Additionally, investigating the long-term impact of participatory planning models, such as community-led restoration projects, could provide actionable insights for policymakers. Sanandaj's hills offer a unique opportunity to redefine urban sustainability in hilly contexts. By prioritizing participatory planning, ecological restoration, and social justice, these hills can transcend their role as battlegrounds of human–nature conflict to become ecological–social corridors that inspire sustainable urban futures worldwide.

Acknowledgments Thanks to all the managers, residents of the Sanandaj Hills, investors, and associations for their cooperation in the interviews, and to everyone who takes steps to promote the sustainability of the Sanandaj Hills.

References

Adams, W. M., & Hutton, J. (2007). People, parks and poverty: Political ecology and biodiversity conservation. *Conservation and Society, 5*(2), 147–183.

Agrawal, A., & Redford, K. (2022). Conservation and displacement: An overview. *Conservation Biology, 36*(5), 123–135.

Agyeman, J. (2022). "Just Sustainabilities": From introducing just sustainabilities: Policy, planning, and practice (2013). In *The sustainable urban development reader* (pp. 199–201). Routledge.

Agyeman, J., Bullard, R. D., & Evans, B. (2021). *Just sustainabilities in policy, planning, and practice*. Zed Books.

Alizadeh, H., Bahrami, B., & Farahani, L. M. (2019). Geophysical setting and urban form of a traditional city: Examination of the Kurdish city of Sanandaj, Iran. *Proceedings of the Institution of Civil Engineers-Urban Design and Planning, 172*(6), 237–247.

Alizadeh, H. (2021). *Building a traditional Kurdish City: The urban morphology of Sanandaj*. Springer Nature.

Alizadeh, H., Doostvandi, M., & Zandsalimi, M. (2024). Redefining sustainable urban development: A critical analysis in the wake of COVID-19. *Sustainable Development, 32*(6), 6163–6181. https://doi.org/10.1002/sd.3018

Anderson, J., Smith, K., & Brown, R. (2015). Qualitative analysis in environmental management. *Environmental Research Letters, 10*(3), 034012. https://doi.org/10.1088/1748-9326/10/3/034012

Anguelovski, I., Shi, L., Chu, E., Gallagher, D., Goh, K., Lamb, Z., Reeve, K., & Teicher, H. (2016). Equity impacts of urban land use planning for climate adaptation: Critical perspectives from the global north and south. *Journal of Planning Education and Research, 36*(3), 333–348.

Bahrami, B., Bahrainy, S. H., & Diba, D. (2014). Spatial organization of public space in hilly settlements, case study: Sanandaj Hills. *Armanshahr Architecture & Urban Development, 6*(11), 35–48. (In Persian).

Bahrami, B., Irandoost, K., & Alizadeh, H. (2019). Spatial planning for hills of Sanandaj: Organizing green and open spaces, creating public spaces, and considering the natural landscape (in Persian). https://doi.org/10.13140/RG.2.2.16904.03841

Bawa, K. S., Kress, W. J., Nadkarni, N. M., & Lele, S. (2004). Beyond paradise: Meeting the challenges in tropical biology in the 21st century. *Biotropica, 36*(4), 437–446.
Bayat, A. (2013). *Life as politics: How ordinary people change the Middle East*. Stanford University Press.
Berkes, F. (2012). *Sacred ecology* (3rd ed.). Routledge.
Berkes, F. (2018). Indigenous knowledge and resource management systems in the Canadian subarctic. *Ecology and Society, 23*(2), 34.
Blomley, N. (2021). The territorialization of property in urban space. *Progress in Human Geography, 45*(4), 769–787.
Braun, V., & Clarke, V. (2006). Using thematic analysis in psychology. *Qualitative Research in Psychology, 3*(2), 77–101. https://doi.org/10.1191/1478088706qp063oa
Brenner, N. (2022). Urban revolution and the critique of neoliberal urbanisation. *Urban Studies, 59*(2), 340–358.
Brenner, N., & Schmid, C. (2015). Towards a new epistemology of the urban? *City, 19*(2–3), 151–182. https://doi.org/10.1080/13604813.2015.1014712
Brown, R., Patel, N., & Gomez, L. (2022). Political economy of urban spaces: A critical review. *Progress in Human Geography, 46*(3), 567–584.
Castree, N. (2021). Nature and society: A dialectical approach to sustainability. *Geoforum, 125*, 885–895.
Chen, H., Liu, W., & Zhao, Y. (2020). Green technologies in urban placemaking. *Landscape and Urban Planning, 202*, 103875.
Creswell, J. W., & Plano Clark, V. L. (2018). *Designing and conducting mixed methods research* (3rd ed.). SAGE Publications.
D'Angelo, M., Rossi, P., & Conti, L. (2022). Capital expansion and ecological degradation in hilly regions. *Environmental Management, 70*(6), 1015–1030.
Davis, M. (2006). *Planet of slums*. Verso Books.
Elmqvist, T., Fragkias, M., Goodness, J., et al. (2013). *Urbanization, biodiversity and ecosystem services: Challenges and opportunities*. Springer. https://doi.org/10.1007/978-94-007-7088-1
Escobar, A. (2020). Pluriversal politics: The real and the possible. *Geoforum, 110*, 184–192.
FAO. (2023). *Global soil erosion trends: 2023 update*. Food and Agriculture Organization of the United Nations. https://www.fao.org/documents/soil-erosion-2023
Folke, C., Biggs, R., Norström, A. V., & Reyers, B. (2023). Social-ecological resilience: New perspectives. *Annual Review of Environment and Resources, 48*, 231–256.
Folke, C., Hahn, T., Olsson, P., & Norberg, J. (2005). Adaptive governance of social-ecological systems. *Annual Review of Environment and Resources, 30*, 441–473. https://doi.org/10.1146/annurev.energy.30.050504.144511
Gupta, M., & Rao, K. (2023). Power structures and urban development in sloped terrains. *Urban Geography, 44*(6), 1123–1145.
Gupta, S., Sharma, R., & Kumar, P. (2023). Neoliberal urbanization and environmental impacts in hilly terrains. *Geoforum, 141*, 103–114.
Harvey, D. (2018). Urbanization under capitalism: Uneven development and the role of capital. *International Journal of Urban Sciences, 24*(3), 215–232.
Harvey, D. (2023). *Spaces of capital: Towards a critical geography* (2nd ed.). Verso Books.
Healey, P. (2022). *Collaborative planning in the 21st century*. Routledge.
Heynen, N. (2019). Urban political ecology: A critical retrospective. *Urban Geography, 40*(7), 943–962.
Huddleston, B., Ataman, E., & Salvatore, M. (2023). Population dynamics in mountainous and hilly regions: A global overview. *Mountain Research and Development, 43*(1), 10–18.
Jones, M., Brown, K., & Ellis, R. (2020). Sustainability in urban planning: A global perspective. *Cities, 105*, 102834.
Kahn, P. H., Jr., & Kellert, S. R. (2002). *Children and nature: Psychological, sociocultural, and evolutionary investigations*. MIT Press.
Kaur, H., & Garg, P. (2023). Urban sustainability assessment tool for hillside planning, design, and development. *Journal of Urban Planning and Development-ASCE, 149*(2).
Kumar, R., Singh, A., & Patel, S. (2022). Indigenous knowledge and sustainable land management in hilly areas. *Journal of Environmental Management, 315*, 115123.

Lefebvre, H. (2021). *The production of space* (Trans. D. Nicholson-Smith). Wiley-Blackwell. (Original work published 1974).

Longley, P. A., Goodchild, M. F., Maguire, D. J., & Rhind, D. W. (2015). *Geographic information systems and science* (4th ed.). Wiley.

Moore, J. W. (2021). *Capitalism in the web of life: Ecology and the accumulation of capital.* Verso Books.

Nagendra, H., Bai, X., Brondizio, E. S., & Lwasa, S. (2018). The urban south and the predicament of global sustainability. *Nature Sustainability, 1*(7), 341–349.

Peck, J., & Theodore, N. (2023). Neoliberal urbanism revisited: Cities and markets in the 21st century. *Urban Studies, 60*(4), 789–805.

Pierce, J., Williams, O. R., & Martin, D. G. (2022). Urban sustainability and the politics of place: A critical review. *Urban Studies, 59*(8), 1557–1574. https://doi.org/10.1177/00420980211064902

Ramirez, J., Perez, M., & Lopez, R. (2021). Participatory GIS for sustainable placemaking in mountainous regions. *Landscape and Urban Planning, 210*, 665–678.

Rees, W. E., Wackernagel, M., & Testemale, P. (2014). Urban ecological footprints: Why cities cannot be sustainable. *Environmental Impact Assessment Review, 14*(4), 223–248. https://doi.org/10.1016/0195-9255(94)90003-5

Reid, H., Evans, B., & Turner, S. (2019). Sustainable urbanism: Lessons from global cities. *Journal of Planning Education and Research, 39*(4), 412–425.

Reij, C., & Garrity, D. (2016). Scaling up farmer-managed natural regeneration in Africa to restore degraded landscapes. *Biotropica, 48*(6), 834–843.

Roy, A. (2011). Slumdog cities: Rethinking subaltern urbanism. *International Journal of Urban and Regional Research, 35*(2), 223–238.

Sandercock, L. (2021). Towards a planning imagination for the 21st century. *Journal of the American Planning Association, 87*(4), 445–456. https://doi.org/10.1080/01944363.2021.1928527

Saunders, B., Sim, J., Kingstone, T., Baker, S., Waterfield, J., Bartlam, B., Burroughs, H., & Jinks, C. (2018). Saturation in qualitative research: Exploring its conceptualization and operationalization. *Quality & Quantity, 52*(4), 1893–1907. https://doi.org/10.1007/s11135-017-0574-8

Schlosberg, D. (2022). *Environmental justice and the new pluralism.* Oxford University Press.

Smith, N. (2023). *Uneven development: Nature, capital, and space* (4th ed.). University of Georgia Press.

Soja, E. W. (2010). *Seeking spatial justice*. University of Minnesota Press.

Switzer, A., Bertolani, M., & Brundiers, K. (2019). Urban hills as socio-ecological systems: Lessons from Lisbon. *Landscape and Urban Planning, 188*, 104–112.

Swyngedouw, E. (2022). Political ecology and the urban question. *Antipode, 54*(5), 1345–1362.

Taylor, A., Smith, J., & Davies, P. (2021). Placemaking principles in urban design. *Urban Studies, 58*(9), 1876–1892.

Tsing, A. L. (2023). *The mushroom at the end of the world: On possibility in capitalist ruins.* Princeton University Press.

UN DESA. (2022). *World urbanization prospects: 2022 revision.* United Nations Department of Economic and Social Affairs. https://www.un.org/development/desa/publications/world-urbanization-prospects-2022.html

UN-Habitat. (2024). *Urbanization and land degradation: Global trends 2024.* United Nations Human Settlements Programme. https://unhabitat.org/global-trends-2024

Walker, J., Mitchell, R., & Platt, S. (2019). Urban ecology and resilience: A review. *Environmental Science & Policy, 98*, 45–53.

World Bank. (2022). *Infrastructure and ecological impacts in hilly regions: 2022 report.* World Bank Group. https://www.worldbank.org/en/topic/infrastructure/hilly-regions-2022

Yu, K.-j., Li, D.-h., & Ji, Q.-p. (2001). Ecological design for landscape and city: Concepts and principles.*Journal of Chinese Landscape Architecture, 6*.

Zhang, L., Chen, P., & Hui, F. (2023). Urban resilience in hilly regions: A global analysis. *Cities, 134*, 104156.

Zimmerer, K. S. (2011). The landscape of agrobiodiversity and the future of land use in the Andes. *Journal of Latin American Geography, 10*(2), 133–157.

Edge Ecologies: Coasts and Deserts

Nurturing Coastal Cities Through Blue Carbon: Nature-Based Solutions for Climate Resilience and Inclusive Development

Junchi Fang, Manqi Yang, and Huixin Huang

Abstract As climate change intensifies, coastal cities are increasingly vulnerable to ecological degradation, extreme weather events, and socio-economic disruption. This chapter explores how nature-based solutions (NBS), particularly those centred on blue carbon ecosystems such as mangroves, seagrasses, and macroalgae, can support multidimensional urban resilience in coastal tourism cities. Using Cancún, Mexico, as a representative case, the chapter identifies three interrelated challenges—ecological vulnerability, economic dependence, and fragmented governance—and proposes a strategic framework that integrates ecological restoration, economic transition, and community participation. To structure this framework, the chapter introduces three innovative blue carbon zones: dynamic coastal buffers, stable ecological anchors, and urban micro-wetlands, each targeting different resilience functions. Three international case studies (Tokyo Bay, Pulicat Lake, and Arcata Wetlands) illustrate diverse NBS implementation models, from top-down engineered approaches to grassroots, community-led initiatives. These cases inform a hybrid strategy tailored to Cancún: dynamic carbon sinks such as Sargassum buffers for coastal protection, stable carbon sinks like mangrove reforestation for water quality and long-term sequestration, and urban-scale micro-wetlands for public engagement. The chapter argues that blue carbon-based NBS can act not only as a climate mitigation tool but as an integrated green solution that links environmental recovery with inclusive urban development and climate justice.

Keywords Climate change · Nature-based solutions · Blue carbon ecosystems · Coastal tourism cities · Resilience governance · Ecological restoration · community participation

J. Fang (✉)
College of Agriculture and Life Sciences, Cornell University, Ithaca, NY, USA

M. Yang
Faculty of Social and Behavioral Science, Universiteit van Amsterdam, Amsterdam, Netherlands

H. Huang
College of Environmental Design, University of California, Berkeley, Berkeley, CA, USA

A. Cheshmehzangi et al. (eds.), *Designing Nature into Cities*, Urban Sustainability, https://doi.org/10.1007/978-981-95-9382-8_8

1 Nature-Based Solutions at the Intersection of Ecology and Equity

Climate change-induced extreme weather events and climate-related disasters, along with marine ecological pollution, have triggered transregional ecological chain reactions, posing severe challenges to both the ecological and economic environments of coastal regions. For instance, in 2017, Hurricane Irma in the Atlantic caused strong winds and storm surges that severely damaged the infrastructure of coastal tourist cities such as Miami and Tampa. The disaster resulted in 134 deaths, forced the evacuation of over 6.5 million people in Florida, and led to an estimated economic loss of approximately $50 billion across the hurricane's path (Gesenhues, 2017). In 2022, extreme rainfall and flooding along Australia's Gold Coast resulted in at least ten deaths and significant beach erosion along the eastern shoreline, severely impacting local tourism. Floodwater also caused sewage overflows, triggering ecological disruption and threats to drinking water safety.

The increasing frequency of coastal climate-related disasters underscores a global imperative to restore the resilience of marine ecosystems and coastal regions, particularly in coastal tourism cities that heavily depend on unique marine ecological resources for resource distribution and industrial development. Extensive coastal urban development practices indicate that these issues arise from both natural and anthropogenic factors and face three primary categories of challenges (environmental-ecological, economic, and sociocultural) (Fig. 1).

The disasters disappointingly affect the low-income and marginalised communities, which not only suffer more severe consequences but also have fewer resources for recovery (Meerow et al., 2019). Contemporary research on coastal urban resilience increasingly focuses on social equity; however, resilience in practice planning often becomes a technical tool of sustaining the order of neoliberalism, reinforcing unequal urban development logics, which even fails to engage with the structural transformations at institutional level (MacKinnon & Derickson, 2012). At the same time, a growing number of green governance-related policies such as ecological restoration have been mobilised as legitimising instruments for sustainable development, reflecting a broader process of green neoliberalism that undermines the local right of participation (Swyngedouw, 2014). As an alternative approach that aims to restore ecosystems while simultaneously achieving social and economic co-benefits, nature-based solution (NBS) approach has been introduced (IUCN, 2020).

Under the framework of NBS, coastal blue carbon ecological system (such as mangrove or seagrass) has increasingly been recognised as an important resource due to their significant carbon sequestration capacity and coastal protection function (Murdiyarso et al., 2015). However, current research and planning policies surrounding the blue carbon resources remain largely focused on its development of carbon market mechanisms (Locatelli et al., 2014; Wylie et al., 2016), with limited attention paid to local embeddedness and implications for socio-spatial justice at urban scale. Especially in the coastal cities driven by tourism and facing high spatial pressures, where questions remain about how coastal blue carbon restoration can be

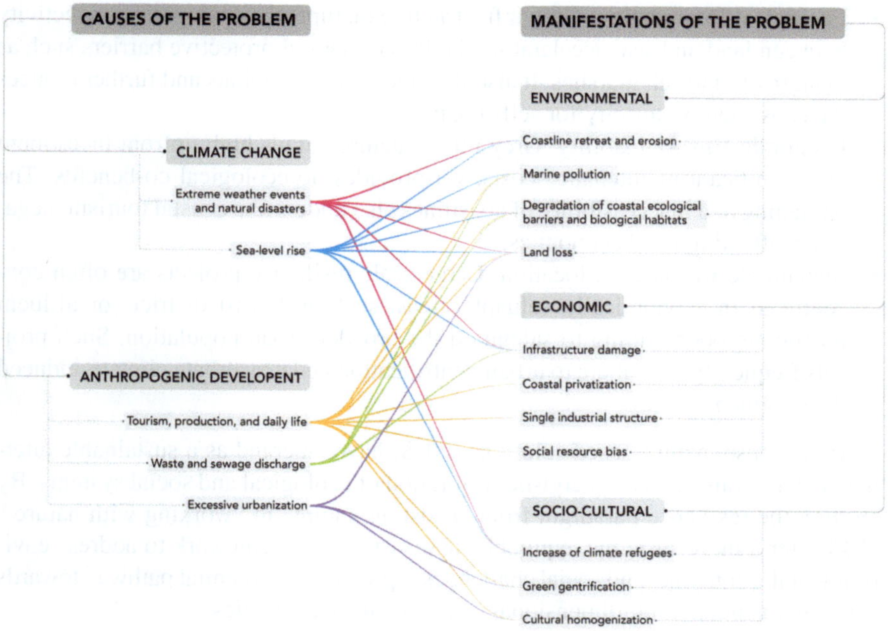

Fig. 1 Causes and manifestations of the problems of coastal tourism cities (Drawn by the authors)

integrated with urban development and effectively incorporate multiple stakeholders in more inclusive and equitable ways.

Therefore, this chapter mainly focuses on coastal blue carbon resources as the core pathway within nature-based solution framework, exploring its potential on localising governance in coastal tourism cities. It aims to develop an integrated resilience governance framework anchored in ecological restoration, grounded in community participation and oriented towards economic transition.

2 Framing the Transition to Blue Carbon Nature-Based Solutions in Coastal Cities

2.1 Challenges and Deficiencies in Resilience Infrastructure Implementation in Coastal Cities

Coastal infrastructure today remains predominantly dominated by grey infrastructure strategies, such as large-scale land reclamation, the construction of breakwaters, seawalls, and sluice gates, which are intended to defend urban areas against rising sea levels (Xu et al., 2025). However, these engineering-driven grey practices often lead to a series of challenges, including ecological degradation, inequitable resource distribution, and long-term economic unsustainability.

- Ecological Degradation: Grey infrastructure disrupts the ecological connectivity between land and sea, accelerating the loss of natural protective barriers such as mangroves and salt marshes. It also damages coastal habitats and further reduces the ecosystem's capacity for self-repair.
- Economic Unsustainability: Grey infrastructure entails high upfront investment and significant maintenance costs, yet provides no ecological co-benefits. The hardening and inaccessibility of coastlines also undermine coastal tourism, negatively affecting local economies.
- Inequitable Resource Allocation: Large-scale resilience projects are often concentrated in economically valuable areas, such as tourist districts or affluent neighbourhoods, failing to safeguard the broader urban population. Such projects frequently contribute to urban gentrification and exacerbate climate-induced displacement.

In response, nature-based solutions (NBS) have emerged as a sustainable alternative that promotes the synergistic restoration of ecological and social systems. By shifting the resilience paradigm from "resisting nature" to "working with nature", NBS offer a more systemic, equitable, and sustainable framework to address environmental, economic, and social challenges—providing a potential pathway towards achieving genuine multidimensional resilience in coastal cities.

2.2 Definition of Nature-Based Solutions (NBS) and the Role of Blue Carbon in NBS Systems

In 2016, the International Union for Conservation of Nature (IUCN) defined NBS as "actions to protect, sustainably manage and restore natural or modified ecosystems, that address societal challenges effectively and adaptively, simultaneously providing human well-being and biodiversity benefits" (International Union for Conservation of Nature [IUCN], n.d.). In 2020, the IUCN released the Global Standard for nature-based solutions (IUCN, 2020), outlining eight core criteria that characterise effective NBS initiatives, which increasingly emphasise the importance of "empowering stakeholders through inclusive participatory governance", "coordinating primary objectives with the continued delivery of ecological and social co-benefits", "adapting to changing conditions through evidence-based and flexible management", and "being sustainably mainstreamed within appropriate legal and policy frameworks".

In summary, NBS should be ecologically grounded and aligned with the long-term evolution of ecosystems, responding adaptively to systemic challenges across multiple scales. Such approaches aim to achieve synergistic outcomes across environmental, social, and economic domains. In the context of the global goal of achieving carbon neutrality by 2050, as proposed by the United Nations, NBS are increasingly recognised as viable strategies for decarbonisation—particularly through the protection, restoration, and sustainable management of natural carbon sinks and reservoirs. Among these, blue carbon ecosystems—such as mangroves, salt marshes, and seagrass meadows—represent some of the most promising NBS pathways in marine and coastal contexts. Blue carbon plays a key role in advancing both land–ocean conservation and climate mitigation agendas, forming a regenerative cycle that supports ecosystem restoration while addressing global warming.

2.3 *Opportunities and Significances of Developing Blue Carbon Resources*

Blue carbon refers to the carbon dioxide absorbed and stored by coastal and marine ecosystems (Crooks et al., 2018). These ecosystems not only serve as vital tools in addressing climate change but also present new opportunities for the economic transformation of coastal cities. As a key ecological restoration approach, the development of blue carbon resources—such as the establishment of algae farms and the restoration of mangroves—can mitigate the impacts of climate change on coastal ecosystems, reduce shoreline erosion, and enhance the adaptive capacity of urban infrastructure. According to the 2019 Special Report on the Ocean and Cryosphere by the IPCC, blue carbon ecosystems like mangroves, seagrass beds, and salt marshes have a carbon sequestration capacity that far exceeds that of terrestrial forests, with the potential to absorb 1–3% of annual global carbon emissions. By advancing the blue carbon economy, coastal cities can contribute to global carbon neutrality while creating new industrial opportunities that promote sustainable economic transformation. In this context, blue carbon has emerged as a nature-based solution (NBS) to address both ecological and developmental challenges in coastal regions.

Ecological Benefits: Restoration measures such as mangrove rehabilitation, seagrass bed reconstruction, and salt marsh management can buffer storm surges, curb coastal erosion, and mitigate sea-level rise. These efforts significantly enhance coastal carbon sequestration, support biodiversity, and increase the overall resilience of interconnected urban-marine ecosystems.

Economic Benefits: Blue carbon development fosters the growth of ecotourism, environmental education, and scientific research industries. It also promotes the development of sustainable biomaterials and supports community-based economies. Furthermore, the establishment of blue carbon credit markets can attract external investment and accelerate industrial transition.

Sociocultural Benefits: Blue carbon initiatives help preserve local marine cultural heritage, restore public coastal spaces to counter the negative effects of shoreline privatisation, and create green job opportunities that reduce climate-induced migration and alleviate social pressure.

Ultimately, blue carbon bridges carbon sink ecosystems, green economic models, and mechanisms for social equity. It is not only an environmental restoration strategy, but also an integrated green solution that holistically addresses the complex ecological, economic, and social challenges faced by coastal regions today.

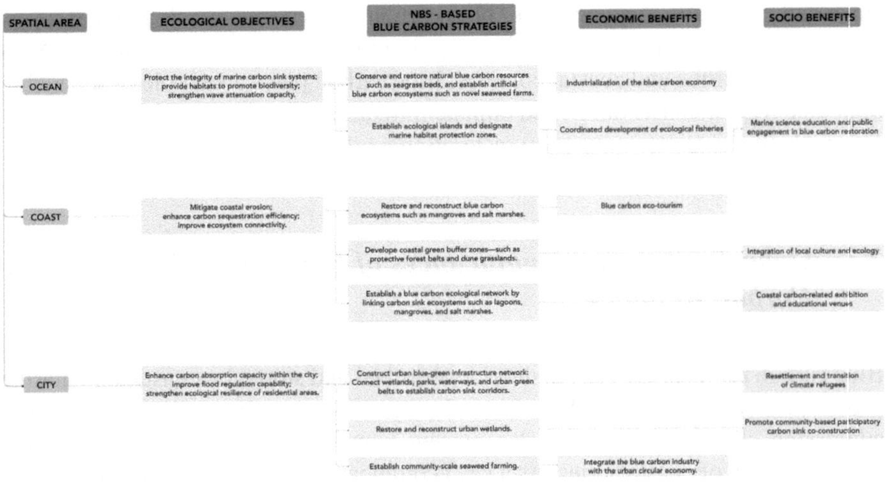

Fig. 2 NBS strategies for three spatial types in coastal cities and their economic and social benefits (Drawn by the authors)

The blue carbon management framework in Figure 2 provides an integrated view of how differentiated strategies across multiple spatial scales can jointly contribute to the ecological conservation, blue carbon development and broader socio-economic outcomes.

2.4 A Resilience Governance Paradigm Rooted in Ecological Restoration, Driven by Economic Transition, and Centred on Community Engagement

Extensive practical experience has shown that despite facing escalating coastal risks and possessing abundant blue carbon resources such as salt marshes, macroalgae, and mangroves, most coastal cities have yet to fully realise their development potential. This chapter aims to establish a resilience enhancement framework for coastal cities that integrates ecological, economic, and sociocultural systems through a nature-based, blue carbon strategy spanning from the ocean to the coastline and inland areas. The proposed framework serves as a model of compound urban resilience—anchored in ecological restoration, mediated by nature, and driven by community co-governance—to help climate-vulnerable coastal cities enhance their adaptive capacity and foster deeper integration of sustainability and social equity.

2.5 Governance Challenges and Legal Integration of Blue Carbon

The ecological and socio-economic potential of blue carbon is deeply influenced by urban planning and legal frameworks. In many coastal cities, blue carbon ecosystems such as mangroves, salt marshes, and seagrass beds are under the jurisdiction of different departments such as marine, forestry, agriculture, and urban planning, resulting in problems such as duplicated supervision and unclear responsibilities in the utilisation of blue carbon resources. The core issue is that currently these blue carbon ecosystems are difficult to be formally included in carbon accounting or land use planning systems, and legal management systems lack consideration for the functions and potential of blue carbon. Meanwhile, the uncertainty of its property rights has also hindered the participation of private investors and coastal communities.

To truly promote blue carbon NBS on a large scale, a more flexible and coordinated governance system is needed. National and local governments, research institutions, and communities should strengthen communication and collaboration, combining national climate goals with local level planning and management. At the same time, it is necessary to develop corresponding legal tools, clarify the definition of blue carbon assets, establish responsible and long-term management departments, and establish civil education institutions. Only by truly incorporating blue carbon ecosystems into legal and institutional frameworks can they become a key force in enhancing climate resilience in cities and coastal areas.

3 Integrated Challenges of Ecology, Economy, and Governance in Cancún

3.1 Commonalities That Exist as Typical Coastal Tourist Cities and Global Coastal Cities

As a typical coastal tourist city, Cancún not only faces the ecological degradation, economic dependence, and social governance challenges common to other coastal tourist cities but also has a geographic characteristic that makes it a highly intertwined urban ecological vulnerability system that faces great threats to its sustainable development.

3.2 Ecological and Economic Background of Cancún

3.2.1 Challenges of Natural Ecological System in Cancún

Geographically, Cancún is situated on the Caribbean coast of the Yucatán Peninsula, a region influenced by a combination of ocean currents, seawater temperature fluctuations, and monsoon patterns, creating an area that is highly susceptible to large-scale Sargassum aggregations (Durand et al., 2024). The massive accumulation of Sargassum will encroach upon the habitats of mangroves and benthic organisms, disrupting photosynthesis and reducing seafloor biodiversity. Furthermore, as the Sargassum washes up and decays on the beach, its absorbed heavy metals are released, contaminating nearshore ecosystems and directly impacting the coastal tourism resources that Cancún relies on for its survival (Chávez et al., 2020).

It is worth noting that the ecological vulnerability of Cancún, a city highly dependent on the tourism industry, is further exacerbated by anthropogenic factors. To maximise the value of the seascape and expand real estate development along the coast, local developers have extensively reclaimed land, significantly encroaching on mangroves and wetlands that would otherwise serve as natural buffer zones. The absence of these ecosystems, which play a critical role in filtering pollutants, regulating water flow, and shielding against extreme weather conditions, also directly diminishes the city's resilience in the face of climate hazards such as hurricanes and flooding.

In addition to its ecological vulnerabilities, Cancún also faces significant hydrological challenges as a coastal city. Because the Yucatan Peninsula is a typical karst landscape, surface water bodies are extremely scarce, so Cancún relies heavily on underground water sources to meet its urban water needs. However, due to the rapid population growth and the unlimited expansion of tourism, water demand in Cancún has surged, leading to the unsustainable extraction of groundwater. At the same time, the development of the service-led urban economy has outpaced the planning and implementation of supporting infrastructure and environmental protection. As a result, Cancún faces not only groundwater shortages but also severe water pollution, which poses risks to public health through contaminated drinking water and threatens the overall health of the urban ecosystem.

3.2.2 Challenge of Social Economic Development in Cancún

Cancún is a city with a hospitality industry heavily reliant on short-term tourist expenditure. Other sectors such as manufacturing and agriculture remain underdeveloped and lack effective economic diversification, leaving the city's economy highly dependent on external markets and natural resources (Wilson, 2008). The tourism industry is dominated by multinational hotel chains and real estate investors, with a high proportion of international tourists, resulting in limited local participation and revenue leakage to global markets.

3.2.3 Challenges of Urban Social Ecological Governance System in Cancún

Cancún, which has transformed from a small fishing village into a renowned international tourist city in just 50 years, has experienced an exceptionally rapid pace of urban expansion. Due to long-standing deficiencies in urban planning and governance, Cancún exhibits severe spatial and social fragmentation, with rapid development in the Hotel Zone contrasting sharply with lagging infrastructure in the city proper (Córdoba Azcárate, 2020). The rapid expansion of the Hotel Zone has outpaced urban infrastructure, worsening issues in water supply, electricity, waste management, and transport. The government's emphasis on tourism has led to a weak cultural sector and overreliance on natural resources, making Cancún's tourism highly vulnerable to environmental threats like Sargassum pollution (López-Contreras et al., 2022).

The city's natural ecological systems are under increasing pressure from both environmental pressures like Sargassum and human activities like land reclamation and overextraction of water. Its tourism-centric economic model lacks diversification and local embeddedness, leading to high external dependence and socio-economic inequalities. Meanwhile, rapid urban expansion driven by the tourism economic benefits has overwhelmed infrastructure and outpaced urban planning, produced spatial segregation, and weakened the overall resilience of the city. In conclusion, the development trajectory of Cancún reflects the deep conflicts between ecological pressures, economic fragility, and governance fragmentation. Similar challenges are faced in different regions; the following three cases examine diverse governance experiences and offer comparative insights into more adaptive and resilient models.

4 Case Study

4.1 Background: Global Implementation of Nature-Based Solutions in Coastal Cities

Nature-based solutions (NBS) have been implemented globally, particularly in coastal cities vulnerable to climate change. Many nations have adopted adaptive strategies to enhance human and ecological resilience along coastlines by focusing on blue carbon ecosystems, including mangroves, wetlands, and algae, which play a dual role in carbon sequestration and ecological restoration. Governments and non-profit organisations in countries such as Japan and Indonesia are working with communities to integrate blue carbon systems into urban ecological planning. For cities whose survival relies heavily on coastal resources—such as Cancún—these global initiatives offer valuable insights across ecological, economic, and governance dimensions.

4.2 *Case Study I: Tokyo Bay, Japan—Artificial Island Construction and Sediment Improvement*

4.2.1 Hydrological Changes in Tokyo Bay: History, Nature, and Development

Tokyo Bay is one of Japan's most densely populated and economically developed coastal regions. Since the 1950s, the area has experienced rapid urbanisation, with a surge in fisheries and logistics industries. The coastal zone underwent significant transformation: shorelines were hardened, and nearshore waters were dominated by informal fisheries, leading to irreversible degradation of water quality.

In 2003, the Japanese government launched the *Tokyo Bay Renaissance Plan*, initiating comprehensive ecological enhancement efforts. Restoration projects targeted tidal flats, estuarine shallows, natural shorelines, reclaimed lands, and salt wedge zones, covering average shoreline lengths of 8 km each.

4.2.2 Government Policy and Public Participation

The *Tokyo Bay Renaissance Plan* was primarily a top-down initiative led by central government agencies. Despite its centralised nature, the project achieved significant ecological improvements within a decade. Its success depended on cross-sectoral coordination among departments of fisheries, environment, transport, and tourism—ensuring that ecological restoration, port functions, and fisheries livelihoods could coexist without conflict.

This model provides a useful precedent for Cancún, where ecological concerns are often overshadowed by tourism capital and where centralised urban planning may offer a more effective means of implementing long-term, site-specific interventions.

To balance policy directives with local economic needs, the Japanese government institutionalised seasonal fishing regulations and formalised previously informal fishing communities into registered committees. As the ecological restoration strategies in Tokyo Bay summarized in Table 1, a public–private co-governance mechanism emerged, with governmental, academic, and community stakeholders jointly managing ecological and economic activities. This participatory governance approach improved both environmental sustainability and economic viability.

Table 1 Ecological restoration strategies and local industry integration in Tokyo Bay

Coastal area	Natural condition	Ecological restoration strategy	Supported industry type	Project outcomes
Sanmaizu	Estuarine sediment zone	Artificial island construction	Coastal recreational services	Habitat restoration; wave and sediment control
Obitsu River Mouth, Banzu Tidal Flat	Tidal flats	Minimal-intervention habitat protection	Small-scale fisheries and bird watching	Seasonal harvest support; bird habitat preservation
Odaiba Seaside Park	Reclaimed coastline	Waterfront park and ferry terminal development	Urban tourism and environmental education	Marine awareness promotion; public space activation

4.3 Case Study II: Pulicat Lagoon, India—Community-Led Mangrove Restoration

4.3.1 Mangrove History: Degradation, Livelihood Loss, and Present Challenges

Pulicat Lake, India's second-largest brackish water lagoon, has long supported local fishing communities whose livelihoods depend on mangrove-based ecosystems. Historically, during colonial rule, the area became a strategic port for Dutch trade, leading to the construction of fortifications and the widespread deforestation of mangroves. These historical disruptions severely degraded the region's ecology and undermined the socio-economic foundations of local communities. Today, ecological instability continues to threaten the sustainability of their traditional livelihood systems.

4.3.2 NGO and Community Collaboration in Mangrove Habitat Restoration

Recognising the ecological and economic importance of mangroves, local residents have gradually adopted the principle that "sustainable fishery depends on mangrove preservation". Since 2012, the community has collaborated with the non-profit CReNIEO on a bottom-up mangrove restoration initiative, supported by global conservation funding.

This project reframes ecological restoration not as a top-down engineering solution but as a collaborative, iterative process led by community expertise. Residents were employed to plant and maintain mangrove zones, which not only restored ecosystem functions but also created local jobs.

Fig. 3 Community-based mangrove restoration process in Pulicat Lagoon *(I. Establishment of Rhizophora Mucronata nurseries; II. Transplantation along canal banks; III. Promotion of natural regeneration through propagule dispersal; IV. Resulting ecological and socio-economic benefits.)* (Drawn by the authors)

Facing environmental stressors—including thermal pollution from nearby factories—the project relied on local ecological knowledge to identify species compatible with the lagoon's varied salinity and soil conditions. Mangrove replanting sites were carefully located to avoid interfering with traditional fishing routes and mudflat access, ensuring ecological and economic zones could coexist. Through 3–5 years of adaptive experimentation, the programme developed a replicable planting protocol and a resilient economic model. As shown in Figure 3, this community-based restoration visualized how community collaboration supported both ecosystem recovery and local livelihood development.

4.4 *Case Study III: Arcata Wetlands, USA—Transforming Infrastructure into Ecological Assets*

4.4.1 Civic Resistance and the Ecological Opportunity for Wastewater Treatment

Located in northern California, the small city of Arcata is situated near Arcata Bay and surrounded by protected coastal redwood forests. In the 1970s, proposals to expand industrial development along the coast, including new wastewater treatment plants, sparked community opposition. Local citizens and university scholars

Fig. 4 Transformation of the Arcata Wetlands over time (1989, 2009, 2024). *Historical aerial imagery showing the progressive ecological restoration of the Arcata coastal area. Source: Google Earth*

organised to transform the required infrastructure into an ecological asset: a publicly accessible wastewater treatment wetland (Fig. 4). As shown in Fig. 4, the Arcata Wetlands underwent a gradual spatial transformation between 1989 and 2024, illustrating how wastewater treatment infrastructure was reconfigured into an ecological asset.

4.4.2 Ecological Wastewater Treatment as Integrated Infrastructure

The Arcata Marsh system, now a key component of the city's wastewater infrastructure, integrates plants, algae, bacteria, and natural water flows to reduce the use of chemicals and prevent marine pollution. This ecological approach not only reduced municipal costs but also created a unique public amenity that supports environmental education and tourism.

What was once a rejected site for waste disposal is now an exemplary case of ecological engineering, where blue carbon processes help transform urban waste into ecological value. For Cancún, this presents a potential model for transforming ecological challenges into tourism-oriented, environmentally conscious economic opportunities (Fig. 5). As shown in Fig. 5, the Arcata Ecological Wastewater Treatment System is organized into three sequential sections, through which wastewater processing is integrated with ecological functions and public landscape provision.

5 Discussion and Implications

Among the current natural solutions to address ecological degradation and climate vulnerability in coastal cities, blue carbon ecosystems are gradually becoming an important medium for integrating ecological restoration, economic restructuring,

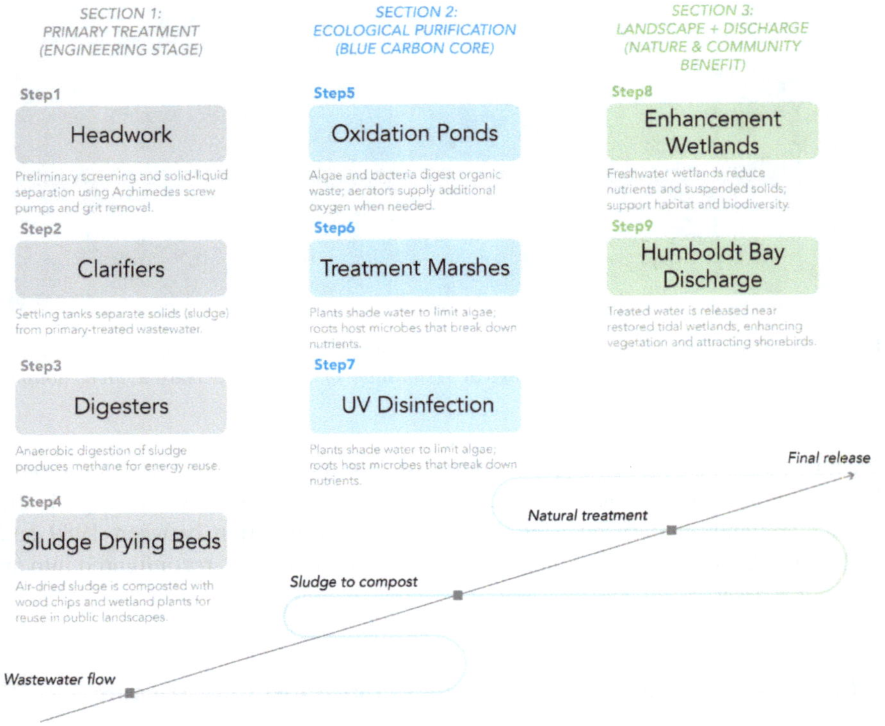

Fig. 5 Process flow diagram of the Arcata Ecological Wastewater Treatment System (*A simplified flow chart illustrating the sequential treatment stages using natural processes—plants, algae, bacteria, and water flow—in the Arcata Marsh system*). (Drawn by the authors)

and social governance due to their highly efficient carbon sequestration capacity and synergistic ecological regulation potential. Blue carbon systems vary in carbon uptake dynamics and storage stability: fast-growing systems like Sargassum offer rapid but short-term sequestration, while mangroves provide long-term carbon storage and ecological buffering. Though micro-wetlands and urban green spaces have limited carbon capacity, they play a vital role in ecological restoration, public awareness, and community engagement.

This chapter categorises blue carbon into three functional types based on their temporal dynamics, spatial scales, and socio-ecological roles. Stable carbon sinks refer to systems such as mangroves and constructed wetlands that provide long-term carbon storage and ecological buffering with minimal fluctuation, serving as foundational infrastructures for coastal resilience. Dynamic carbon sinks, represented by organisms like Sargassum and seaweeds, are characterised by their rapid carbon uptake and capacity for seasonal or event-based transformation. These systems often participate in closed-loop processes connecting ecological restoration with green industries, such as algae harvesting or oyster aquaculture. The third category, termed socially embedded carbon sinks, includes micro-scale and

Table 2 Blue carbon typologies: ecological roles and urban applications (Drawn by authors)

Blue carbon type	Definition	Functions	Examples	Benefits
Dynamic carbon sinks	Fast-growing biomass systems with short-term but high-efficiency carbon uptake	Rapid carbon absorption, support regenerative cycles	Sargassum, seaweed farms, algae-based systems	Circular economy potential, water purification, resource reuse
Stable carbon sinks	Long-term, low-variability carbon storage systems with ecological buffering	Sequester carbon over decades, stabilise coastal habitats	Mangroves, salt marshes, constructed wetlands	Coastal protection, biodiversity support, long-term carbon storage
Social embedded carbon sinks	Small-scale, community-based systems that integrate nature into urban life	Connect people with ecosystems, promote decentralised stewardship	Community wetlands, green yards, micro-gardens	Enhances public engagement, urban cooling, education, local ecological awareness

community-managed spaces—such as neighbourhood wetlands and green yards—that act as public interfaces between human activities and ecological processes. These systems integrate blue carbon storage into everyday urban life, enhancing public engagement, environmental awareness, and decentralised stewardship. By distinguishing these three categories, the framework helps to clarify how different blue carbon strategies contribute synergistically to urban climate resilience as summarized in Table 2.

5.1 Coastal Nearshore Areas: Ecological Buffer Strategies Based on Dynamic Carbon Sinks to Reduce the Sargassum Hazard (Ecological Restoration-Based)

Frequent and large-scale Sargassum outbreaks have had a severe impact on beach landscapes and marine ecosystems, resulting in both ecological degradation and economic stress. The accumulation of Sargassum obscures underwater light, leading to a decrease in the productivity of the benthic ecosystem, while the release of ammonia, hydrogen sulphide, and heavy metal pollutants during the decomposition process exacerbates the deterioration of the water quality, resulting in the formation of brown tides. To cope with this situation, ecological buffer strategies based on the dynamic blue carbon mechanism should be adopted in the nearshore coastal area. The deployment of floating ecological facilities or low-intervention biobased interception systems can effectively attenuate wave energy in shallow nearshore waters, reduce Sargassum landing intensity and shoreline erosion, and protect the structural stability of sandy beaches. In addition, the interception system can actively collect floating Sargassum to prevent its onshore accumulation and subsequent decomposition. It also opens up possibilities for nearshore aquaculture, such as oyster farming, and reduces surface algal cover near the coast, improving light availability for coral reefs and seagrass beds.

5.2 Lagoon Transition Zone: Using Stabilised Carbon Sinks to Promote Ecological Restoration and the Establishment of a Reuse Zone for Green Resources (Economic Transition-Oriented)

As an important ecological buffer between the coast and the city, Cancún's inner lagoon originally carried the functions of water filtration, biodiversity maintenance, and carbon storage. However, driven by urban landfill, overfishing, and tourism development, a large number of mangrove systems in the area have been damaged, and the ecological structure has been fractured, leading to increased eutrophication and pollution in the water body, weakening its ecological functions and reducing the city's resilience to marine disasters (Alongi, 2002). Therefore, the reconstruction of the mangrove system should be prioritised, with establishment of the lagoon ecological restoration unit that provides long-term carbon sequestration.

A finger-like mangrove belt can be established in the wetland area, adopting the method of Arcata Wetlands in the United States, leveraging the synergistic effects of mangroves, Sargassum, microorganisms, and water body to reduce the content of chemical pollutants in the water flowing into the lagoon (Yu, 2020). A suspended vertical oyster cultivation zone can be constructed in the nearshore area as a low-intervention, ecologically synergistic bio-utilisation method. Sargassum intercepted by the finger-like mangrove belt can be collected and processed through artificial wetland facilities into green recycled materials such as biomass fuel or sustainable building materials (Córdoba Azcárate, 2020; López-Contreras et al., 2022). At the same time, some of the Sargassum will be introduced to the community for artificial treatment and provide raw materials for community-oriented industrial production (Case study iii: Arcata Ecological Wastewater Treatment System). Within this system, blue carbon plays a dual role: on the one hand, it serves as a stable natural carbon sink in the lagoon area, assuming the role of an ecological barrier in the periphery of the city; on the other hand, it promotes the transformation of resources and community participation by means of the green material cycle path, realising the triple win-win situation of ecological restoration, economic compensation, and social connection.

5.3 Urban Inland Node Space: Building Community Carbon Sink Nodes to Realise Socially Embedded Blue Carbon Governance Space for Public Carbon Sinks

The ecological function of urban space has long been overlooked in urban interiors, especially in low-lying areas inhabited by marginalised communities and characterised by inadequate infrastructure. By constructing micro-wetlands and vegetated green corridors in the low-lying areas, community spaces, and along major inland roads, parts of the urban interior are integrated into a "public carbon sink" system, where residents plant locally adapted mangrove saplings and aquatic vegetation to build a network of socially embedded blue carbon micro-systems, so as to incorporate carbon sinks into the spatial details of everyday life. Through topographic modification and vegetation planning, these interventions form public corner spaces with ecological regulating capacity, where blue carbon serves not only as a medium for carbon storage and system regulation, but also as a tool for ecological governance. Table 3 sets out the detailed strategy framework across coastal, intertidal, and urban inland spaces, linking spatial conditions with functional priorities, blue carbon orientations, and corresponding interventions. In this way, the concept of blue carbon extends beyond the coastline to become part of daily routines and community engagement, turning micro-scale green spaces into critical nodes for fostering urban multidimensional resilience.

6 Conclusion

This chapter proposes a pathway framework for the use of blue carbon ecosystems as a multidimensional resilience governance tool, centred on the sustainable transformation of coastal tourism cities in the context of climate change. The case study of Cancún shows that when the blue carbon space is divided into three functional areas, namely "dynamic carbon sink", "stable carbon sink", and "socially embedded carbon sink", it not only contributes to ecological restoration and carbon sink enhancement, but also serves as a catalyst for economic transformation, community participation, and institutional reshaping.

Through the comparative analysis of the three scenarios, this chapter demonstrates the practical logic of blue carbon ecosystems under different governance mechanisms. These experiences point to the fact that in addressing complex urban climate vulnerability, blue carbon should not only be viewed as an ecological restoration tool for carbon offsetting alone but also be understood as a comprehensive resilient restoration medium across scales, systems, and subjects. Applying this thinking to Cancún will also help to promote new pathways for embedding "nature-based solutions" into urban socio-ecological system governance while providing actionable spatial zoning ideas and governance strategies at the practical level. The restoration and protection of blue carbon ecosystems often crosses the boundaries

Table 3 Detailed strategy framework (drawn by authors)

Spatial	Functional focus	Blue carbon oriented	Detail strategy	Example of blue carbon strategy
Coastal area	Ecological restoration	Blue Carbon Buffer Zone	Artificial reefs buffer wave energy	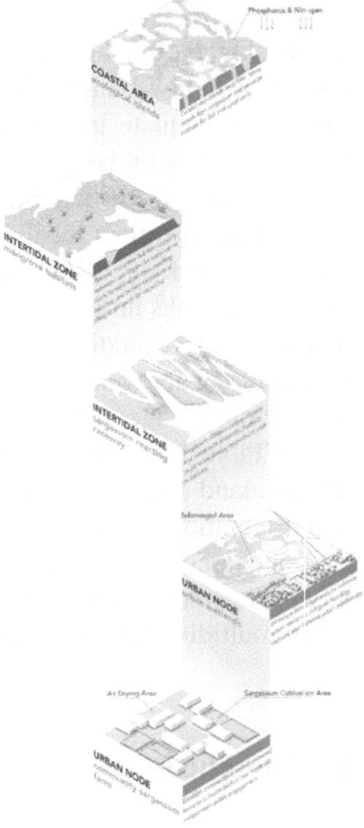
			Artificial reefs intercept Sargassum and help reduce algal bloom intensity	
Intertidal zone of the Lagoon	Economic transformation (Transition of Ecological Industries)	Productive Blue Carbon Belt	A finger-like mangrove system is introduced between the tourist zone and residential areas, serving as an ecological buffer	
			Artificial filtration modules are used to purify water between the coastline and urban settlements	
			A floating oyster farming area is established near the urban residential zone	
			The collected Sargassum is processed into recycled materials such as biomass fuel	
Urban inland node space	Community participation	Social Carbon Node	Ecological water-retention parks are developed in low-lying urban areas through to extend the wetland system	
			Mangrove seedlings can be planted to enhance purification capacity and improve local biodiversity	
			Micro ecological corridors, linear green spaces, and drainage ditches are established along major urban roads	
			Water is filtered and channelled into the subsurface to replenish groundwater	
			Urban open spaces are converted into micro-gardens using recycled Sargassum for water retention and plant cultivation	

of multiple sectors and stakeholders (e.g. water authorities, environmental protection bureaus, tourism bureaus, and private landowners), and it is difficult to sustain or prioritise the maintenance of blue carbon systems into policy agendas if there is not a clear legal status, management authority, and financial mechanism, which could be further emphasised by the institutional embedding of blue carbon. Institutional embedding of blue carbon can be further emphasised in future research to transform the blue carbon system into a "formal object" in the urban governance system. In the face of increasing climate challenges, blue carbon is not only about environmental restoration, but also an important medium for realising synergistic urban social justice and ecological governance.

References

Alongi, D. M. (2002). Present state and future of the world's mangrove forests. *Environmental Conservation, 29*(3), 331–349. https://doi.org/10.1017/S0376892902000231

Chávez, V., Uribe-Martínez, A., Cuevas, E., Rodríguez-Martínez, R. E., van Tussenbroek, B. I., Francisco, V., Estévez, M., Celis, L. B., Monroy-Velázquez, L. V., Leal-Bautista, R., Álvarez-Filip, L., García-Sánchez, M., Masia, L., & Silva, R. (2020). Massive influx of pelagic sargassum spp. on the coasts of the Mexican Caribbean 2014–2020: Challenges and opportunities. *Water, 12*(10), Article 10. https://doi.org/10.3390/w12102908

Córdoba Azcárate, M. (2020). *Stuck with tourism: Space, power, and labor in contemporary Yucatan*. University of California Press.

Crooks, S., Windham-Myers, L., & Troxler, T. (2018). Defining blue carbon: The emergence of a climate context for coastal carbon dynamics. In L. Windham-Myers, S. Crooks, & T. Troxler (Eds.), *A blue carbon primer: The state of coastal wetland carbon science, practice and policy* (pp. 1–8). CRC Press. https://doi.org/10.1201/9780429435362

Durand, L., Sundberg, J., & Rodríguez-Martínez, R. (2024). Seaweed blooms in paradise: Ecological reflexivity, governance and the sargassum crisis in the Mexican Caribbean. *Ocean and Coastal Research, 72*. https://doi.org/10.1590/2675-2824072.23089

Gesenhues, A. (2017, December 13). *Hurricane Irma was the No. 1 top trending Google search in the US & globally for 2017*. Search Engine Land. https://searchengineland.com/hurricane-irma-no-1-top-trending-google-search-us-globally-2017-288127

International Union for Conservation of Nature. (n.d.). *Nature-based solutions*. https://iucn.org/our-work/nature-based-solutions

International Union for Conservation of Nature. (2020). *IUCN global standard for nature-based solutions: A user-friendly framework for the verification, design and scaling up of NbS* (1st ed.). IUCN. https://doi.org/10.2305/IUCN.CH.2020.08.en

Locatelli, T., Binet, T., Kairo, J.G. et al. (2014). Turning the Tide: How Blue Carbon and Payments for Ecosystem Services (PES) Might Help Save Mangrove Forests. AMBIO 43, 981–995. https://doi.org/.proxy.uba.uva.nl/10.1007/s13280-014-0530-y

López-Contreras, A. M., Valenzuela, P. N., García, B. C., Driegen, J., Lwanga, E. H., Domin, P., Gurrola, M. P., Rosas-Luis, R., Gómez, Y. V., & de Vrije, T. (2022). *Sargassum in mexico: From environmental problem to valuable resource*. https://doi.org/10.18174/574423

MacKinnon, D., & Derickson, K. D. (2012). "From Resilience to Resourcefulness: A Critique of Resilience Policy and Activism," Papers in Evolutionary Economic Geography (PEEG) 1212, Utrecht University, Department of Human Geography and Spatial Planning, Group Economic Geography, revised Jun 2012.

Meerow, S., Pajouhesh, P., & Miller, T. R. (2019). Social equity in urban resilience planning. *Local Environment, 24*(9), 793–808. https://doi.org/.proxy.uba.uva.nl/10.1080/13549839.2019.1645103

Murdiyarso, D., Purbopuspito, J., Kauffman, J. et al. (2015). The potential of Indonesian mangrove forests for global climate change mitigation. Nature Clim Change 5, 1089–1092. https://doi.org/10.1038/nclimate2734

Swyngedouw, E. (2014). Where is the political? Insurgent mobilisations and the incipient "return of the political." *Space and Polity, 18*(2), 122–136. https://doi.org/.proxy.uba.uva.nl/10.1080/13562576.2013.879774

Wilson, T. (2008). Economic and social impacts of tourism in Mexico. *Latin American Perspectives - LAT AMER PERSPECT, 35*, 37–52. https://doi.org/10.1177/0094582X08315758

Wylie, L., Sutton-Grier, A. E., & Moore, A. (2016). Keys to successful blue carbon projects: Lessons learned from global case studies. *Marine Policy, 65*, 76–84. https://doi.org/10.1016/j.marpol.2015.12.020

Xu, X., O'Sullivan, J. J., Abolfathi, S., Keenahan, J., Pearson, J. M., & Salauddin, M. (2025). Advances in understanding the challenges and opportunities of hybrid sea defence approaches for coastal resilience. *Environmental Challenges, 100130.* https://doi.org/10.1016/j.envc.2025.101130

Yu, K. (2020). *Sanya Mangrove Park: Restoration of a lifeless landfill into an ecological mangrove park.* Turenscape. https://www.turenscape.com/en/project/detail/4677.html

Nature-Based Solutions for Climate-Resilient Coastal Shorelines: Strategic Lessons for Kuakata, Bangladesh

Tasnim Ferdous and Monisha Momtaz

Abstract Climate change and natural resource exploitation pose unprecedented challenges, threatening the environment, socio-economic stability, and the livelihoods of millions in a densely populated nation like Bangladesh. As its coastal region grapples with these escalating challenges, there is a growing recognition of the multifaceted benefits that nature-based solutions (NbS) offer in fostering resilience and sustainability. With its low-lying topography, susceptibility to natural disasters, and gradual environmental degradation, a dynamic coastal city like Kuakata is exceptionally vulnerable to sea-level rise, soil erosion, salinization of agricultural land and drinking water, and infrastructural damage. The chapter comprehensively analyzes the wide array of strategies, robust policies through case studies, and qualitative data collection and explores possible NbS interventions at the coastal shorelines of Kuakata. The case studies from regions with similar geophysical nature help explore potential transferable lessons for Kuakata. NbS implementation in the area can potentially mitigate the impacts of climate change and align with broader sustainable development goals, including poverty reduction, food security, and biodiversity conservation. The findings highlight pivotal foundations including community participation, hybrid approaches, and policy integration in ensuring the efficacy of these lessons and underscore the strategies of tailoring NbS to local contexts. In summary, this chapter attempts to extensively research the versatility of nature-based approaches in relevant case studies and harness strategical insights for a climate-resilient framework to stabilize the shoreline of Kuakata in Bangladesh.

Keywords Nature-based solutions (NbS) · Coastal shorelines · Climate change · Climate-resilient development · Sustainability

---

T. Ferdous
University of Manitoba, Winnipeg, MB, Canada

M. Momtaz (✉)
Department of Architecture, Bangladesh University, Dhaka, Bangladesh
e-mail: monisha20053@bu.edu.bd

A. Cheshmehzangi et al. (eds.), *Designing Nature into Cities*, Urban Sustainability, https://doi.org/10.1007/978-981-95-9382-8_9

1 Introduction

Climate change is a pervasive and escalating threat that transcends national borders, affecting ecosystems, economies, and societies globally. Driven by anthropogenic activities, burning of fossil fuels, and deforestation, the concentration of greenhouse gases in the atmosphere is increasing. Rising temperatures, melting ice, rising sea levels, and increasing frequency of extreme weather events are some of the key global impacts of climate change. These changes disrupt natural habitats, threaten food security, exacerbate water scarcity, and lead to severe socio-economic consequences, particularly for vulnerable populations. As the impacts of climate change intensify, the need for a comprehensive and localized understanding of these effects becomes increasingly critical (IPCC, 2022). As global cities seek strategies to adapt to these challenges while maintaining ecological integrity, nature-based solutions (NbS) have emerged as a promising paradigm. NbS leverage natural systems and ecosystem processes to address socio-environmental challenges, aiming to enhance biodiversity, provide ecosystem services, and improve community resilience (IUCN, 2020).

The concept of NbS acknowledges that, when harnessed wisely, nature provides a repertoire of resilient and sustainable mechanisms to address the impacts of climate change (Fernandes et al. 2022). This chapter aims to explore the application of NbS in the context of coastal resilience through a comparative analysis of case studies from three different countries—Indonesia, Vietnam, and Bangladesh and connect these insights to the context of Kuakata, a vulnerable coastal city in Bangladesh.

The rationale for this chapter lies in the urgent need to strengthen climate resilience in cities like Kuakata through holistic, evidence-based, and community-centered approaches. By analyzing successful NbS interventions through case studies, we identify pathways that can be adapted and implemented in Kuakata and similar coastal regions. This chapter integrates cross-regional experiences to inform practical recommendations for NbS design and implementation, emphasizing community participation, ecosystem functionality, and policy integration. By synthesizing theory, practice, relevant case studies, and place-based insights, this chapter contributes to the growing body of knowledge on nature-based climate resilience and seeks to offer actionable frameworks for a vulnerable coastal city like Kuakata.

2 Background Knowledge

2.1 Understanding Nature-Based Solutions (NbS)

Nature-based solutions (NbS) are defined by the International Union for Conservation of Nature (IUCN) as "*actions to protect, sustainably manage, and restore natural or modified ecosystems that address societal challenges effectively and adaptively, simultaneously providing human well-being and biodiversity benefits*" (IUCN,

2020). This framework promotes an integrated approach to urban and environmental resilience that leverages ecological processes to mitigate climate risks while enhancing ecosystem services.

NbS can include a range of interventions, from reforestation, wetland restoration, and green urban infrastructure to hybrid solutions that combine engineered and natural systems, water resource management, or protected area management.

The accelerating impacts of climate change—rising sea levels, extreme heat, intensified storms, and erratic rainfall patterns pose serious threats to urban and coastal systems globally. NbS provide a sustainable alternative or complement to conventional "gray" infrastructure by:

- Mitigating heat through urban greening and tree canopies
- Enhancing flood management through natural drainage and wetland systems
- Reducing coastal erosion and storm surge damage through mangroves and coral reefs
- Supporting carbon sequestration, biodiversity conservation, and improved air and water quality (Cohen-Shacham et al., 2016) (Fig. 1)

Urban NbS, such as green roofs, vegetated swales, and urban forests, can also deliver co-benefits including health, recreation, and social cohesion (Mukherjee et al., 2022, p. 25). In coastal regions, NbS can restore the protective function of natural buffers, like mangrove afforestation, wetland restoration, and agroecological farming to increase adaptive capacity, against climate-induced hazards such as cyclones, floods, and erosion (Tasnim et al., 2020), and contribute to community-led disaster preparedness.

To maximize impact and ensure sustainability, the IUCN has proposed eight guiding principles for successful NbS interventions (Andrade et al., 2020), which include:

- Addressing societal challenges (e.g., climate adaptation, disaster risk reduction)
- Designing for biodiversity and ecosystem integrity
- Implementing at landscape scale
- Promoting inclusive governance and stakeholder engagement
- Balancing trade-offs and ensuring equitable benefits

Fig. 1 Mangrove benefits & ecosystem services (Source: Authors)

- Employing adaptive management and evidence-based monitoring
- Sustainable financing and institutional integration
- Mainstreaming into policy, planning, and legal frameworks

Applying these principles can support the achievement of long-term climate and development goals. Furthermore, NbS align with multiple global policy frameworks, including the Paris Agreement on climate change, the Sendai Framework for Disaster Risk Reduction, and the UN Sustainable Development Goals (SDGs), particularly Goals 11 (Sustainable Cities and Communities), 13 (Climate Action), and 15 (Life on Land). By integrating ecosystem services into planning and policy, NbS offer a pathway for building adaptive capacity while ensuring biodiversity conservation and inclusive development (Frantzeskaki et al., 2019). Several countries have embedded NbS within their Nationally Determined Contributions (NDCs), such as Vietnam, Indonesia, and Bangladesh, which use NbS for coastal protection, reforestation, and urban resilience (Seddon et al., 2020).

2.2 *Wetlands and Coastal Development Strategies*

Wetlands, including mangroves, tidal marshes, and estuarine systems, play a foundational role in acting as natural buffers, attenuating wave energy, reducing storm surge impacts, and regulating hydrological cycles. In the face of climate change-induced threats such as sea-level rise, cyclonic storms, and salinity intrusion, preserving and restoring wetlands is critical not only for ecological sustainability but also for safeguarding human livelihoods and infrastructure (Barbier et al., 2011; Spalding et al., 2014).

Coastal wetlands serve as multifunctional landscapes that provide a range of ecosystem services. Ecosystem-based adaptation (EbA) refers to the use of biodiversity and ecosystem services as part of an overall strategy to help people adapt to the adverse effects of climate change. In coastal zones, EbA interventions include mangrove afforestation, the restoration of salt marshes and dunes, and the use of natural floodplains and tidal basins for water retention and filtration (Munang et al., 2013). Mangroves, in particular, reduce the height and energy of storm surges through their dense root and canopy structures, with evidence showing that every 100 m of mangrove forest can reduce wave height by up to 66% (Gedan et al., 2011).

In the context of Bangladesh, where the coastal landscape is highly dynamic and prone to erosion, these ecosystems are vital. Projects such as the Coastal Afforestation Program, initiated in the 1960s, have illustrated how planted mangroves along embankments and foreshores can reduce cyclone damage and delay embankment failure (Haque et al., 2019). Similarly, natural dunes and beach vegetation along the Bay of Bengal, though often neglected, contribute to shoreline stabilization and provide an additional line of defense against extreme weather events (Fernández-Montblanc et al., 2020).

2.3 Barriers to the Adoption of Nature-Based Coastal Strategies

Despite their proven benefits, several barriers hinder the adoption of wetland and EbA-based coastal strategies. A major one is the engineering bias in policy and planning. In many developing countries, gray infrastructure—embankments, sluice gates, and sea walls—remains the default response to coastal hazards, often seen as offering faster, more tangible protection (Asare et al., 2023). This bias marginalizes NbS in official plans and budgets. Land tenure issues also pose considerable challenges. In Bangladesh, mangrove forests and mudflats are often classified as state-owned or unallocated, limiting community stewardship. Such ambiguity fosters encroachment, aquaculture expansion, and unsustainable resource use (Iftekhar & Islam, 2004). Land titling rarely reflects wetland value, creating misalignment between development and conservation.

Finally, limited interdisciplinary capacity and data impede NbS design and evaluation. Effective interventions require collaboration among ecologists, engineers, planners, and social scientists, yet institutional silos and scarce expertise block integration. Weak long-term monitoring further constrains assessment of NbS effectiveness.

2.4 Integrating Coastal Wetland Strategies into Local Planning

To overcome these challenges, a paradigm shift is needed from viewing wetlands as wastelands or barriers to development to recognizing them as vital infrastructure for resilience. Mainstreaming wetland-based adaptation into coastal and urban planning can be achieved through inclusive policy frameworks, incentive-based conservation programs, and capacity-building initiatives at the local level (Seddon et al., 2020). In Bangladesh, integrating EbA into the National Adaptation Plan and aligning it with spatial planning at the sub-district level could support more targeted interventions in vulnerable areas like Kuakata (Ministry of Environment, Forest and Climate Change, 2022). Cross-learning from the Mekong and Indonesian cases shows that combining ecological restoration with co-management and tenure security can yield durable outcomes for both people and nature.

3 Site Context

Kuakata, located in the Patuakhali district of southwestern Bangladesh, is a critical area within the Ganges tidal floodplain. (Fig. 2). It falls within the coastal zone of Bangladesh which comprises 19 sea-facing districts and nearly 600 km of coastline (Uddin et al., 2023). Residents of this city frequently encounter cyclones,

Fig. 2 Location of Kuakata in the map of Bangladesh (Source: Authors)

particularly in May and November (Nasreen et al. 2023). It is geographically unique due to its flat coastal topography, intersecting river systems, and direct exposure to the Bay of Bengal. While it is renowned for its scenic beauty and cultural significance, the town faces increasing risks from cyclones, storm surges, salinity intrusion, and waterlogging. These risks are compounded by the degradation of natural buffers such as mangrove forests and inadequate infrastructure planning (Bangladesh Forest Department, 2016).

Sedimentation in this area primarily results from the movement of suspended fine silt and clay particles carried by tidal flows. (Figs 3 and 4). These rivers often possess sandy beds and are bordered by mudflats composed of fine, easily erodible sediments (Bangladesh Water Department Board, 2022). The soil in the region is predominantly non-calcareous clay, with silty and slightly calcareous textures near riverbanks and transitional zones closer to the lower Meghna estuary. According to the Urban Development Directorate, Ministry of Housing and Public Works 2021, notable sediment deposition occurs due to the low elevation and proximity to tidal influences; coastal upazilas in this region face heightened risks from sea-level rise associated with climate change. The city shoreline is located in a high-risk coastal zone of Bangladesh that is particularly vulnerable to a range of natural hazards, including flooding, storm surges, tropical cyclones, salinity intrusion, riverbank erosion, and waterlogging. These vulnerabilities are primarily a consequence of the region's low elevation and geographic exposure to climatic and hydrological extremes. The area frequently experiences severe cyclonic events, which are often accompanied by wind speeds exceeding 220 km/h and storm surges reaching 5–6 m in height (Urban Development Directorate, 2021).

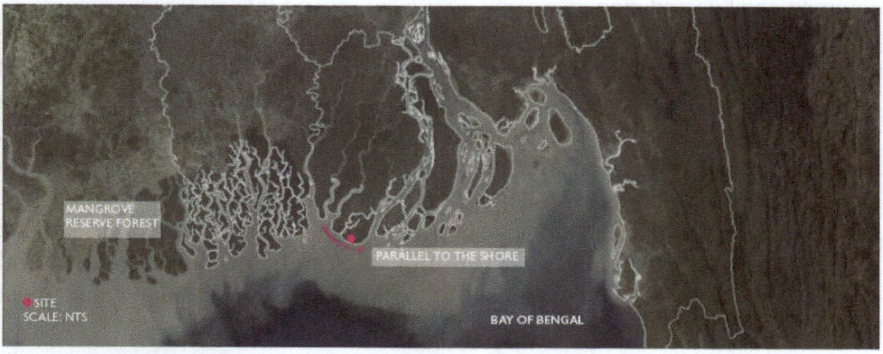

The Longshore currents move parallel to the shore moving sediment along the coast causing erosion and accretion in different areas.

Fig. 3 Direction of longshore current (Source: Authors)

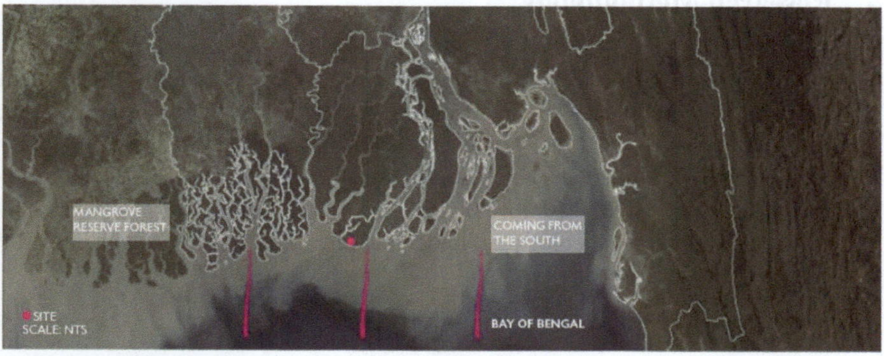

The cross shore currents hitting the coast from the south move sediments towards and away from the shore. They play a significant role during storms.

Fig. 4 Direction of cross shore current (Source: Authors)

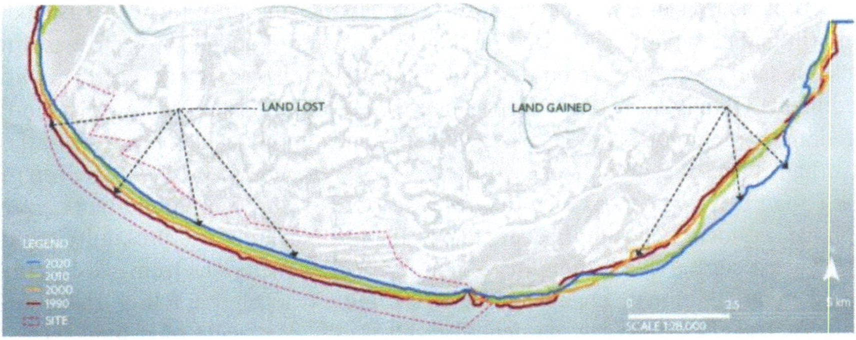

Fig. 5 Changing shorelines, year 1990–2020 (Source: Authors)

Natural coastal features such as mangrove forests, sand dunes, tidal rivers, and creeks serve as critical ecological buffers, mitigating the impact of these hazards by absorbing wave energy, stabilizing sediments, and reducing wind and water velocity. The degradation of these protective ecosystems is contributing to an elevated risk of disasters, making local communities more susceptible to climatic shocks and environmental degradation.

Landsat analyses show significant shoreline change in Kuakata. (Fig 5) The Payra-Kuakata Comprehensive Plan 2021–2041 reported nearly 500 m of coastal erosion between 1988 and 2022 (Urban Development Directorate, 2021). From 1989 to 2020, the southwest coast lost 4.76 km^2, including mangroves, farmland, settlements, and beaches (Aishi & Hasan, 2022). As a beach town, Kuakata attracts industries and tourism dependent on coastal resources, but its dynamic shoreline demands tailored strategies for sustainable management amid climate change and development pressures. Ensuring the careful management of natural and physical resources is essential for both the local community and local authorities.

4 Research Methodology

For this research, Kuakata town was chosen to be the focus because of its current structure plan, geographical unique location, and its potential to grow into a bigger, more tourist centric city. While some approach of NbS integration has already started, the study aims to delve deeper into a more structured, practice-based, localized nature-based approach. Moreover, within the scope of the study, primary observation, photo mapping, and contextual understanding have helped narrow down the study's site to be this coastal city.

4.1 Research Approach and Data Collection

The research primarily applies qualitative methodology, including semi-structured interviews, supported by quantitative data to assess NbS applied to regions with similar geographical context and discusses key insights into implementing them to the site. The research primarily follows two approaches for data collection—interviews and desktop analysis which includes case studies, academic papers, practical work compilations, census reports, government records and by-laws, international organization reports, and Development Project Proforma (DPP). The findings have been analyzed through GIS mapping, timelines, scope of community participation, green and infrastructure restoration, and livelihood benefits.

4.2 Desktop Analysis and Interviews

Multiple case studies are taken purposively from South Asian sub-tropical regions with similar climatic conditions, soil type, geographical nature, and socio-economic values. Case studies from Vietnam, Indonesia, and Bangladesh share similar climatic environmental challenges as well as flood-prone coastal regions that have demonstrated effective NbS approaches. The projects from Bangladesh provide execution strategies, local challenges, and feasibility that strengthen possibilities of applying transferable strategies to another city in the same country. The similarity in livelihood, geographical and climatic context, and development trajectory of all three countries allow room for transferable, adaptable applications of the lessons learnt from the case studies.

Semi-structured, open-ended interviews were conducted with experts working in the relevant fields for qualitative data collection as well as accessible resources. The interviews were conducted online and documented through notes and recordings. Due to the experts' availability, accessibility, and overall time constraints, this study interviewed 11 experts (Table 1).

The participants in the interviews were asked a combination of technical, social, and institutional inquiries to reflect on their first-hand experiences in implementing NbS projects. Questions were designed to explore their primary focuses during the interventions, such as testing soil quality or water retention and movement. To better understand the social aspects, community participation and local knowledge exchanges were also discussed along with any bureaucratic constraints, funding, or administrative challenges they might have faced during project execution. The interviews sought lessons and best practices from first-hand experiences to effectively strategize NbS efforts for Kuakata.

Table 1 Interview participants and affiliation (Source: Authors)

Participant code	Area of expertise/designation	Affiliation type/sector[a]
P1	Environment, Climate Change and Research System	Independent Consultant
P2	Geography and Environment	Education
P3	Physical Geography, GIS and Remote Sensing, Marine and Spatial Planning	Researcher
P4	Disaster Management	Education
P5	Landscape Architecture	Independent Consultant
P6	Disaster Management	Education
P7	Landscape Architecture, Planner, Climate Adaptation	Professional Association, International NGO
P8	Water Resources, Hydrologic and Morpho-dynamic Modelling	Researcher
P9	Architecture and Planning	Professional Association
P10	Landscape Architecture, Academic Research	Education
P11	Program Management, Environmental Leadership	International NGO

[a]Affiliation type/sector have been generalized to maintain the anonymity of the participants

4.3 *Visual Assessment*

An integrated visual assessment was carried out through a combination of photo documentation as well as GIS-based mapping. Sectional sketches and drawings were done to understand locations of crop lands and rural or urban settlements in relation to the coastline. Landsat analyses with the aid of GIS-driven data were conducted to map visible changes in the shorelines (Fig. 5), which also helped identify areas of significant land loss. The observations and analysis were recorded through graphical representations, tables, and flowcharts.

5 Relevant Case Study Analysis

5.1 *Sustainable Landscape Management Program (SLMP), Central Java, Indonesia*

Indonesia's northern coastline, particularly in Central Java, has experienced significant degradation due to mangrove deforestation, unsustainable aquaculture practices, and the construction of rigid coastal infrastructure. The "Building with Nature" initiative, implemented between 2015 and 2021, exemplifies an integrated approach to coastal resilience by combining ecological restoration with socio-economic development (Wetlands International, 2022).

A baseline assessment in Demak revealed that over 3 km of land had already been lost to the sea, with entire villages submerged and income losses of up to 60–80% for local communities, underscoring the severity of prior conditions and the urgency for intervention (Wetlands International, 2024).

Key interventions included the dismantling of hard sea walls and the installation of permeable bamboo brushwood structures to facilitate natural sediment deposition, thereby promoting mangrove colonization (Winterwerp et al., 2013). Additionally, the program introduced silvo-fisheries, integrated mangrove-shrimp aquaculture systems that enhanced both ecological and economic outcomes. The key takeaways from this case study

- Restoration of over 200 ha of mangrove habitat
- Approximately 80% reduction in erosion-related displacement
- Improved shrimp yields, leading to increased incomes for local fishers
- Strengthened community governance through the establishment of coastal field schools

Despite these successes, questions remain about long-term sediment dynamics and whether the permeable structures can withstand future sea-level rise and storm intensification (Van Wesenbeeck et al., 2016). On the economic side, while shrimp yields improved, the approach must ensure that aquaculture practices remain ecologically sustainable and do not lead to soil subsidence or water quality degradation over time (Van Wesenbeeck et al., 2016). This case underscores the efficacy of hybrid NbS approaches that synergize natural and engineered systems, emphasizing the importance of community involvement in restoring degraded coastal ecosystems and enhancing livelihoods.

5.2 Mekong Delta Integrated NbS and Wetland Management, Vietnam

The Mekong Delta in Vietnam is highly susceptible to sea-level rise, saltwater intrusion, and storm surges, posing significant risks to its 17 million inhabitants and its role as a major rice-producing region. In response, the Vietnamese government, in collaboration with organizations such as Deutsche Gesellschaft für International Zusammenarbeit (GIZ) and IUCN, implemented a multi-scalar NbS project focusing on coastal mangrove restoration, policy reform, and risk-sensitive land-use planning (GIZ & IUCN, 2017).

The project involved restoring mangroves along sea dykes and integrating agro-ecological zoning to support climate-resilient agriculture. Financial mechanisms, including Payment for Ecosystem Services (PES), were introduced to incentivize community participation in long-term stewardship. The summarized outcomes are as follows:

- Restoration of 9000 ha of mangrove forests across three provinces
- Enhanced protection of over 350 km of coastal dyke systems
- Incorporating NbS to Vietnam's national adaptation and disaster risk reduction plans

Vietnam's 2017 Resolution 120 ("Adapting to Nature") established a strategic policy framework for integrated, climate-resilient development in the Mekong Delta (Pham et al., 2022). From 2015 to 2020, satellite-based assessments revealed that 27,322 ha (about 34.1%) of this increased mangrove cover was due to restoration or new planting across seven coastal provinces, despite losses of 16,138 ha (20.3%) from erosion and land conversion (Tinh et al., 2022).

However, restoration success varied; many projects showed low mangrove survival rates due to poor site selection, lack of long-term monitoring, and insufficient community engagement (Damastuti et al., 2023). Integrated mangrove–shrimp aquaculture (silvo-aquaculture) offers triple-win benefits—ecological restoration, diversified income, and improved livelihoods, but its adoption remains limited by suboptimal design, limited training, and unclear tenure arrangements (McIlveen & Hung, 2019).

This initiative demonstrates how integrating NbS into policy frameworks and providing economic incentives can lead to large-scale successes in coastal resilience, although the sustainability of outcomes depends on strong ecological monitoring, inclusive governance, and equitable benefit-sharing.

5.3 Mangrove Afforestation and Coastal Greenbelt Projects, Bangladesh

Bangladesh, one of the nations most vulnerable to climate change, has undertaken several NbS initiatives for coastal protection, notably the Coastal Afforestation Program and the Greenbelt Project initiated in the 1980s. These projects, led by the Forest Department with support from the UNDP and FAO, focused on planting mangroves and salt-tolerant species along the Bay of Bengal's coastline and riverbanks (Bangladesh Forest Department, 2020).

In recent years, the emphasis has shifted toward community-managed afforestation, embankment protection through vegetation, and promoting adaptive livelihoods such as honey production and crab fattening. The key outcomes of this case are:

- Establishment or rehabilitation of over 200,000 ha of mangrove forests
- Reduction in cyclone-induced damages in afforested areas, as observed during Cyclone Sidr in 2007 and Cyclone Amphan in 2020
- Increased biodiversity, soil stabilization, and salinity regulation
- Formation of Community Forest User Groups (CFUGs) for long-term protection and benefit-sharing

Satellite-based monitoring shows that from 1965 to 2013, coastal afforestation exceeded 209,000 ha, but early monoculture plantations suffered mortality rates up to 80% due to ecological mismatch and weak management (Mahmood et al., 2023). Later programs, such as CBACC-CF and CRPARP, applied multi-species planting and participatory approaches, restoring over 13,800 ha between 2008 and 2016 (Mahmood et al., 2023). Nevertheless, afforestation remains more successful in densely populated floodplain areas, where community engagement and land security are higher. The durability of benefits is threatened when tenure is uncertain or livelihood incentives decline (Mahmood et al., 2023). Moreover, while livelihoods like honey and crab fattening provide diversification, evidence is lacking on equitable income distribution—particularly among landless households. This case shows Bangladesh's shift from monoculture to more resilient, socially inclusive NbS. However, long-term effectiveness hinges on governance structures, tenure clarity, and ensuring equitable benefit-sharing across communities.

5.4 Comparative Assessment

Despite these successes, challenges persist, including land tenure issues, encroachment, and the need for integrated governance across sectors. Nonetheless, these initiatives represent some of the most substantial applications of NbS in South Asia.

As shown in Table 2 and Fig. 6, Indonesia, Vietnam, and Bangladesh demonstrate different scales and approaches to mangrove restoration, yet all achieve tangible benefits. Restoration efforts reinforce coastlines and embankments, diversify livelihoods; from silvo-fisheries to honey and crab production and generate additional income streams such as PES revenues. Community involvement remains central, with co-management structures, field schools, and gender-inclusive participation strengthening both social equity and governance. Together, these outcomes highlight how NbS can simultaneously protect infrastructure, sustain local economies, and build climate-resilient coastlines.

Table 2 Comparative assessment of the case studies (Source: Derived from case studies by authors)

Country	Restored mangrove area (ha)	Coastline protected (km)	Livelihood benefits	Community involvement
Indonesia	200	~20 km of eroding delta coastline reinforced (Tol & Wetlands International, 2022)	Improved shrimp yields, farmers trained, coastal field schools improved farming efficiency	In coastal field schools and silvo-fisheries
Vietnam	9000	350 coastal embankments reinforced	Annual PES revenue; stewardship contracts for households, increased rice yields	Post-restoration contracts, co-management policy incentives
Bangladesh	200,000	~8 km of mangrove belts fronting embankments (Gijón Mancheño et al., 2025)	Increased honey production and crab fattening, increased average household income (22%) (UNFCCC, n.d.)	Marginalized groups involved by CBACC-CF, CFUGs formed; female participation (~42%) (UNFCCC, n.d.)

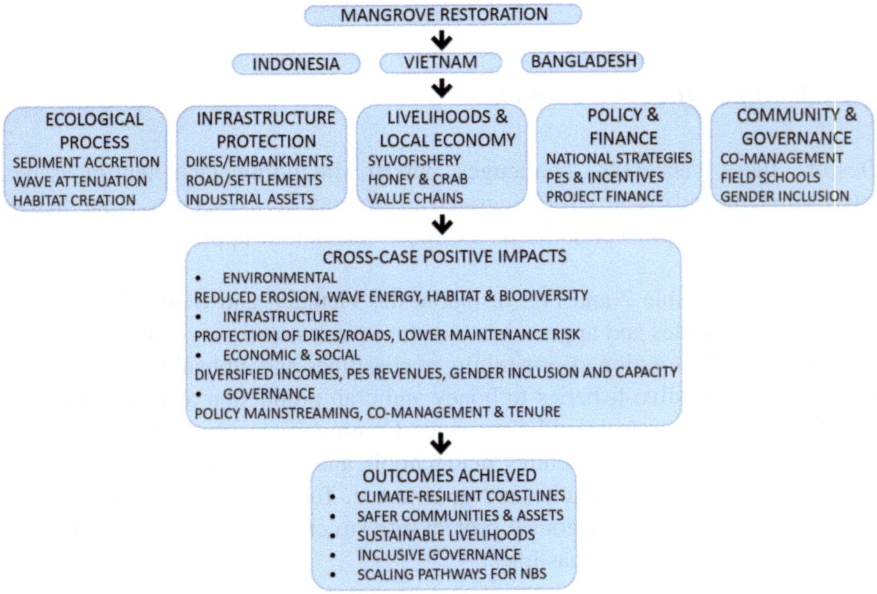

Fig. 6 Flow chart analysis of the case studies (Source: Derived from Case Studies by Authors)

6 Case Study Insights Integrated with Localized NbS Strategy for Kuakata

The aforementioned case studies offer valuable insights for coastal towns like Kuakata, which are exposed to cyclones, tidal floods, and salinity intrusion. Indonesia's hybrid model, bamboo breakwaters with mangrove restoration offers an alternative to heavy reliance on concrete embankments for shoreline erosion control. Vietnam's use of PES and policy integration highlights the need to embed NbS into regional and national development plans. Bangladesh's afforestation programs emphasize that long-term community engagement, livelihood integration, and secure land-use rights are critical for NbS success.

Drawing on these insights, a contextualized NbS strategy for Kuakata should combine enabling pillars with operational strategies, as illustrated in the integrated framework (Fig. 7). The pillars—community engagement, hybrid approaches, and policy integration provide the foundation, while the strategies focus on spatial targeting, hybrid green-gray solutions, participatory governance, and mainstreaming NbS into planning.

- Spatial Identification and Prioritization: Targeting degraded, erosion-prone zones for mangrove regeneration
- Hybrid Green-Gray Infrastructure: Combining mangrove belts with bamboo fencing to stabilize the coastline
- Green Drainage and Retention Ponds: Integrating drainage corridors and retention ponds to manage stormwater and floods
- Participatory Governance: Engaging communities in planning, monitoring, and maintenance
- Mainstreaming NbS: Embedding NbS in local development and disaster risk reduction plans to ensure long-term sustainability

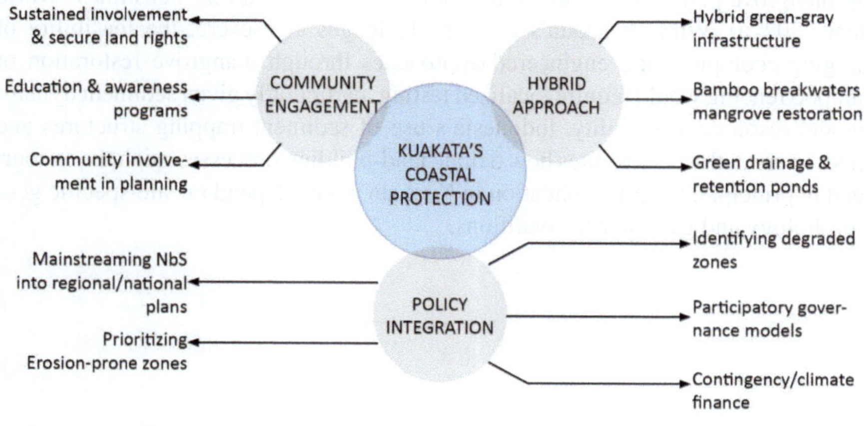

Fig. 7 Approach to Kuakata's coastal protection (Source: Authors)

By aligning these interventions with lessons from other coastal regions, Kuakata can shift from reactive to proactive climate adaptation, offering a replicable model for other polder towns in Bangladesh and beyond. The lessons and understandings from the case studies provide key insights on probable strategy implementation to the coast of Kuakata to protect the shoreline.

6.1 Multiple Lines of Defense

The Vietnamese Mekong Delta, one of the world's most climate-vulnerable regions, has adopted the Multiple Lines of Defense (MLD) approach, which integrates NbS with engineered structures for sustainable coastal protection (World Bank, 2022b). Implemented by WWF, this approach emphasizes urban green infrastructure such as retention ponds, vegetated canals, and tree-lined streets to manage stormwater and mitigate heat (Royal Government of Vietnam, 2019). Approaches such as bioswales, rain gardens, or multipurpose green spaces may hold potential if adapted to local hydrology and social contexts. Community ponds and seasonal drainage channels, if ecologically designed, could inspire flexible strategies for water storage, biodiversity support, and erosion reduction. These insights highlight possibilities for co-benefits such as ecosystem restoration and socio-economic uplift while acknowledging Kuakata's distinct ecological and institutional realities (Mekong Connect Forum, 2024).

6.2 Wetland and Coastal Protection Strategies

The Mangroves for Coastal Resilience (M4CR) project in Indonesia highlights the benefits of restoring tidal connectivity and using hybrid solutions, such as combining mangrove plantations with bamboo breakwaters, to stabilize coastlines (World Bank, 2022a). While Kuakata's erosion challenges are severe, the feasibility of merging ecological and engineered approaches through mangrove restoration or bamboo fencing would require localized testing, particularly given sediment dynamics and resource availability. Indonesia's use of sediment-trapping structures and restored creek flows illustrates how deltaic land-building processes might be encouraged in principle, though application in Kuakata would depend on site-specific geomorphology and governance conditions.

6.3 Community-Driven Resilience

All three case studies emphasize the critical role of community participation in sustaining NbS interventions. In Bangladesh, community-led mangrove restoration initiatives have successfully engaged local populations, particularly women, in managing nurseries, planting, and maintaining mangrove trees (UN Women, 2021). Gender-sensitive engagement models using activities such as rooftop gardening, vegetated drains, or awareness campaigns could be explored, but the precise form should emerge from local priorities and institutional capacity. The Mekong and Indonesian experiences further show that building community stewardship reduces long-term dependence on external aid, an approach Kuakata could adapt under its own cultural and governance frameworks.

6.4 Policy, Planning, and Institutional Mechanisms

Successful NbS in the Mekong Delta and Indonesia depended on coordination across sectors and clear mandates at different government levels (GIZ, 2023). The Payra-Kuakata Comprehensive Plan: 2021–2041 already includes nature-based planning in Union Parishad and Upazila development frameworks (UDD, Ministry of Housing and Public Works, 2021, p. 33). Durable NbS requires clarity, collaboration, and financing. For Kuakata, this may involve cooperation among Local Government Engineering Department (LGED), Department of Forests, the Department of Environment, local NGOs, and academic institutions, but only within existing governance capacity. International cases highlight the importance of contingency funds and climate finance, though Kuakata faces fiscal and bureaucratic limits. Regional monitoring platforms can also serve as models, but would require adaptation to local technological and institutional conditions.

7 Conclusion and Way Forward

This chapter explored the growing relevance and application of nature-based solutions (NbS) in enhancing climate resilience in vulnerable coastal regions, with a particular focus on the case of Kuakata, Bangladesh. Drawing on global and regional case studies from Vietnam's Mekong Delta to Indonesia's mangrove restoration projects and Bangladesh's own community-driven initiatives, the discussion demonstrated how NbS can serve as effective, low-cost, and ecologically integrated approaches to managing the multifaceted challenges posed by climate change in deltaic and coastal areas.

A key takeaway is the adaptability of NbS across diverse ecological and sociopolitical contexts. While the physical features, environmental vulnerabilities, and governance systems vary between the Mekong Delta, Indonesia, and Bangladesh, all three regions have shown that solutions rooted in natural processes, whether through mangrove restoration, integrated wetland systems, or green urban infrastructure, can yield wide-ranging benefits. However, successful implementation hinges on the context sensitivity of interventions. What works in the Mekong Delta with its extensive canal networks and salinity prevention sluices may not directly translate to Kuakata without significant adaptation. Local geomorphology, community capacities, institutional arrangements, and cultural practices must inform every stage of NbS planning and execution.

Kuakata emerges in this chapter as a microcosm of broader coastal climate risks faced by low-lying regions across South and Southeast Asia. Its exposure to cyclones, saline intrusion, tidal flooding, and erosion mirrors the vulnerabilities observed in many other deltaic towns. Development pressure further threatens NbS in Kuakata. Rapid infrastructure growth and unregulated tourism have driven land cover change, dune erosion, and vegetation loss. Lacking safeguards, such activities undermine ecosystems and heighten climate risks (Mostafiz et al., 2021; Sarker, 2024). Yet, these exposures also position Kuakata as a critical site for demonstrating how integrated NbS strategies can be operationalized at the local scale. The synthesis of lessons points to several immediate entry points: hybrid green-gray coastal infrastructure using bamboo fences and mangrove belts, incorporation of green drainage systems into urban planning, and participatory models of governance that place local communities—particularly women and youth, at the heart of resilience-building.

To move forward, a multi-pronged approach is needed. First, investment in pilot projects that test localized NbS models in Kuakata and similar areas should be prioritized. These projects can serve as living laboratories for understanding how NbS perform over time under varying climatic conditions. Second, research institutions must be engaged in developing evidence-based monitoring systems to evaluate ecological, social, and economic outcomes. Data from such initiatives can guide adaptive management and improve the design of future interventions. Third, participatory planning must be institutionalized. This requires building capacity at the grassroots level, creating inclusive decision-making platforms, and ensuring that benefits such as enhanced livelihoods and reduced disaster risks are equitably shared. Regional collaboration, including knowledge exchange between countries implementing NbS, can accelerate learning and foster innovations suited to local realities. Finally, policy coherence and cross-sectoral coordination remain essential. Embedding NbS in local development, disaster risk reduction, and land-use planning frameworks will help mainstream these approaches and secure sustainable financing. Government agencies, NGOs, and donor institutions must align their efforts to create enabling environments for NbS to thrive.

In conclusion, nature-based solutions offer a promising pathway for climate adaptation in coastal towns like Kuakata. While challenges remain, the lessons from global best practices demonstrate that with the right combination of science, community engagement, and policy support, NbS can transform vulnerabilities into opportunities for resilient, inclusive, and sustainable coastal futures.

Acknowledgments Sincere gratitude to CSLA student research grant for supporting this research and Dr. Haseeb Irfanullah, Independent Consultant of Environment, Climate Change & Research System, and Sanjoy Roy, Researcher of Physical Geography, GIS & RS, Marine Spatial Planning, for sharing their knowledge and resources which set the backbone for this study.

References

Aishi, A. F., & Hasan, K. (2022). Time-series analysis of landcover dynamics and their relation with coastline migration along Kuakata coast, Bangladesh using remote sensing techniques. *Geology, Ecology, and Landscapes, 8*(1), 71–85. https://doi.org/10.1080/24749508.2022.2097374

Andrade, A., Cohen-Shacham, E., Dalton, J., Edwards, S., Hessenberger, D., Maginnis, S., Maynard, S., McElwee, P., Murti, R., Nelson, C., Ruiz, V., Siikamäki, J., & Vasseur, L. (2020). *Guidance for using the IUCN global standard for nature-based solutions* (1st ed.). IUCN. https://doi.org/10.2305/IUCN.CH.2020.09.en

Asare, B. A., Boateng, I., & Appeaning Addo, K. (2023). Mapping and assessment of coastal infrastructure for adaptation to coastal erosion along the coast of Ghana. *Anthropocene Coasts, 6*(1), 1–15. https://doi.org/10.1007/s44218-023-00026-6

Bangladesh Forest Department. (2016). *Coastal afforestation and forest management strategies*. Ministry of Environment, Forest and Climate Change.

Bangladesh Forest Department. (2020). *Coastal afforestation and mangrove enrichment plan in Bangladesh: Community-based adaptation to climate change*.

Bangladesh Water Development Board. (2022). *Coastal Embankment Improvement Project, Phase 1 (CEIP-1)*. Ministry of Water Resources.

Barbier, E. B., Hacker, S. D., Kennedy, C., Koch, E. W., Stier, A. C., & Silliman, B. R. (2011). The value of estuarine and coastal ecosystem services. *Ecological Monographs, 81*(2), 169–193. https://doi.org/10.1890/10-1510.1

Cohen-Shacham, E., Walters, G., Janzen, C., & Maginnis, S. (Eds.). (2016). *Nature-based solutions to address global societal challenges*. International Union for Conservation of Nature (IUCN). https://portals.iucn.org/library/sites/library/files/documents/2016-036.pdf

Damastuti, E., Van Wesenbeeck, B., Leemans, R., Groot, R., & Silvius, M. (2023). Effectiveness of community-based mangrove management for coastal protection: A case study from Central Java, Indonesis. *Ocean & Coastal Management, 238*, 106498. https://doi.org/10.1016/j.ocecoaman.2023.106498

Fernández-Montblanc, T., Duo, E., & Ciavola, P. (2020). Dune reconstruction and revegetation as a potential measure to decrease coastal erosion and flooding under extreme storm conditions. *Ocean & Coastal Management, 188*, 105075. https://doi.org/10.1016/j.ocecoaman.2019.105075

Fernandes, R., Panwar, V., & Sen, M. (2022). Nature-based Solutions for urban climate resilience in South Asia: Cases from Bangladesh, India and Nepal. Climate and Development Knowledge Network (CDKN).

Frantzeskaki, N., McPhearson, T., Kendal, D., Bulkeley, H., Dumitru, A., Walsh, C., Noble, K., van Wyk, E., Ordóñez Barona, C., Oke, C., & Pintér, L. (2019). Nature-based solutions for urban climate change adaptation: Linking science, policy, and practice communities for evidence-based decision-making. *BioScience, 69*(6), 455–466. https://doi.org/10.1093/biosci/biz042

Gedan, K. B., Kirwan, M. L., Wolanski, E., Barbier, E. B., & Silliman, B. R. (2011). The present and future role of coastal wetland vegetation in protecting shorelines: Answering recent challenges to the paradigm. *Climatic Change, 106*(1), 7–29. https://doi.org/10.1007/s10584-010-0003-7

Gijón Mancheño, A., Brander, L., Jafino, B. A., Urrutia, I., & Kazi, S. (2025, July 21). *Mangroves for coastal resilience in Bangladesh: Identification of potential locations, assessment of triple dividends, and cost estimation*. World Bank. https://openknowledge.worldbank.org/entities/publication/88b5d4e6-3954-480b-8346-23db01cd55ee

GIZ & IUCN. (2017). *Nature-based solutions for coastal protection in Vietnam*. https://www.iucn.org/resources/publication/2017-nature-based-solutions-vietnam

GIZ. (2023). *Mekong delta climate resilience programme*. https://2023.snrd-asia.org/mekong-delta-climate-resilience-programme/

Haque, M. A., Islam, M. T., & Rahman, M. M. (2019). Coastal afforestation in Bangladesh: Current status and future considerations. *Environmental Challenges, 1*, 100007. https://doi.org/10.1016/j.envc.2020.100007

Iftekhar, M. S., & Islam, M. R. (2004). Managing mangroves in Bangladesh: A strategy analysis. *Journal of Coastal Conservation, 10*(1), 139–146. https://doi.org/10.1652/1400-0350(2004)010[0139:MMIBAS]2.0.CO;2

Intergovernmental Panel on Climate Change. (2022). *Sixth assessment report: Impacts, adaptation, and vulnerability*. https://www.ipcc.ch/report/ar6/wg2/

IUCN. (2020). *Global standard for nature-based solutions*. https://portals.iucn.org/library/node/49070

Mahmood, R., Zhang, L., & Li, G. (2023). Assessing effectiveness of nature-based solution with big earth data: 60 years mangrove plantation program in Bangladesh coast. *Ecological Processes, 12*(11). https://doi.org/10.1186/s13717-023-00419-y

McIlveen, S., & Hung, P. Q. (2019). Improving livelihoods and increasing coastal resilience: A look at integrated mangrove-shrimp aquaculture in Vietnam. *ENA Quarterly, 23*(4).

Mekong Connect Forum. (2024). *Proceedings of the Mekong Connect Forum 2024*. An Giang, Vietnam.

Ministry of Environment, Forest and Climate Change. (2022). *National Adaptation Plan of Bangladesh (2023–2050)*. https://www.undp.org/bangladesh/publications/national-adaptation-plan-bangladesh-2023-2050

Mostafiz, R. B., Islam, M. M., & Rahman, M. M. (2021). Technical and social approaches to study shoreline change of Kuakata, Bangladesh. *Frontiers in Marine Science, 8*, 730984. https://doi.org/10.3389/fmars.2021.730984Frontiers

Mukherjee, M., Wickramasinghe, D., Chowdhooree, I., Chimi, C., Poudel, S., Mishra, B., Ali, Z. F., & Shaw, R. (2022). Nature-based resilience: Experiences of five cities from South Asia. *International Journal of Environmental Research and Public Health, 19*(19), 11846. https://doi.org/10.3390/ijerph191911846

Munang, R., Thiaw, I., Alverson, K., Mumba, M., Liu, J., & Rivington, M. (2013). Climate change and Ecosystem-based Adaptation: A new pragmatic approach to buffering climate change impacts. *Current Opinion in Environmental Sustainability, 5*(1), 67–71. https://doi.org/10.1016/j.cosust.2012.12.001

Nasreen, M., Hossain, K. M., & Khan, M. M. (2023). *Coastal disaster risk management in Bangladesh*. Routledge eBooks. https://doi.org/10.4324/9781003253495

Pham, T. T., Vien, N. N., Vo, Q. T., Tang, T. K. H., Nguyen, T. L., Tran, N. M. H., et al. (2022). *Opportunities and challenges for mangrove restoration in the Mekong delta: Status, policies and stakeholder*. Bogor, Indonesia: CIFOR.

Royal Government of Vietnam. (2019). *Highlights of resolution on climate-resilient development of Mekong Delta*. https://en.baochinhphu.vn/highlights-of-resolution-on-climate-resilient-development-of-mekong-delta-11140608.htm

Sarker, J. U. (2024). Prospects and challenges at Kuakata sea beach for sustainable tourism development: A qualitative analysis. *East West Journal of Business and Social Studies, 12*, 95–116. https://doi.org/10.70527/ewjbss.v12i.144

Seddon, N., Smith, A., Smith, P., Key, I., Chausson, A., & Girardin, C. (2020). Getting the message right on nature-based solutions to climate change. *Global Change Biology, 27*(8), 1518–1546. https://doi.org/10.1111/gcb.15513

Spalding, M. D., McIvor, A., Tonneijck, F. H., Tol, S., & van Eijk, P. (2014). *Mangroves for coastal defence: Guidelines for coastal managers & policy makers*. Wetlands International and The Nature Conservancy.

Tasnim, T., Anzum, F., Md. Irfanullah, H., Seddon, N., & Huq, S. (2020, August). *A roadmap for nature-based solutions in Bangladesh*. https://www.icccad.net/wp-content/uploads/2020/08/ICCCAD_NBSI_Policy_Brief_NbSRoadmap_August2020.pdf

Tinh, P. H., MacKenzie, R. A., Hung, T. D., Vinh, T. V., Ha, H. T., Lam, M. H., Hanh, N. T. H., Tung, N. X., Hai, P. M., & Huyen, B. T. (2022). Mangrove restoration in Vietnamese Mekong Delta during 2015–2020: Achievements and challenges. *Frontiers in Marine Science, 9*, 1043943. https://doi.org/10.3389/fmars.2022.1043943

Tol, S., & Wetlands International. (2022). *Building with nature for safe, prosperous and adaptive coastlines in Indonesia*. PANORAMA Solutions. https://panorama.solutions/en/solution/building-nature-safe-prosperous-and-adaptive-coastlines-indonesia

Uddin, M. J., Niloy, M. N. R., Haque, M. N., & Fayshal, M. A. (2023). Assessing the shoreline dynamics on Kuakata, coastal area of Bangladesh: A GIS- and RS-based approach. *Arab Gulf Journal of Scientific Research, 41*(3), 240–259. https://doi.org/10.1108/AGJSR-07-2022-0114

UN Women. (2021). *Women reap the rewards of planting mangroves in Bangladesh*. https://wrd.unwomen.org/explore/insights/women-reap-rewards-planting-mangroves-bangladesh

United Nations Framework Convention on Climate Change. (n.d.). *Community-based adaptation to climate change through coastal afforestation (CBACC-CF) project – Bangladesh*. https://unfccc.int/climate-action/momentum-for-change/activity-database/community-based-adaptation-to-climate-change-through-coastal-afforestation-cbacc-cf-project-in-bangladesh

Urban Development Directorate, Ministry of Housing and Public Works. (2021). *Payra-Kuakata comprehensive plan focusing on eco-tourism: Structure plan 2021–2041* (p. 33).

Van Wesenbeeck, B., Tonneijck, F., Siry, H., de Vries, J., Cronin, K., van Eijk, P., Wilms, T., & Noor, Y. (2016). Building with nature: Sustainable protection of mangrove coasts. *Terra et Aqua, 6*, 5–10.

Wetlands International. (2022). *Building with Nature in Indonesia: Restoring an eroding coastline and inspiring action at scale (2015–2021)*. https://www.wetlands.org/publication/building-with-nature-in-indonesia-restoring-an-eroding-coastline-and-inspiring-action-at-scale-2015-2021/

Wetlands International. (2024, May 15). *Building with nature Indonesia*. https://www.wetlands.org/case-study/building-with-nature-indonesia/

Winterwerp, J. C., Erftemeijer, P. L. A., Suryadiputra, N., van Eijk, P., & Zhang, L. (2013). Defining eco-morphodynamic requirements for rehabilitating eroding mangrove-mud coasts. *Wetlands, 33*(3), 515–526. https://doi.org/10.1007/s13157-013-0409-x

World Bank. (2022a). *New project will support large-scale mangrove conservation and restoration in Indonesia*. https://www.worldbank.org/en/news/press-release/2022/06/07/

World Bank. (2022b). *Sustainable Landscape Management Project (SLMP) in Indonesia: Final evaluation report*.

Greening a Desert City: A Case Study of Riyadh

Erna Husukic, Jane Handal, and Monica Moscatelli

Abstract With a massive construction and reconstruction drive, the city of Riyadh is staged for an unprecedented transformation led by the ambitious Vision 2030 of the Kingdom of Saudi Arabia initiated on April 25, 2016. This chapter preludes by examining the historical evolution of the city of Riyadh, focusing on the interrelated relationship between urban growth and the natural environment. It identifies key challenges and underlying opportunities in the city's efforts to achieve environmental, economic, and social sustainability. Central to the chapter is a scrutiny of transformative greening initiatives, particularly the Green Riyadh Program and other environmentally responsive urban projects, designed to enhance climate resilience, improve air quality, increase green space per capita, and elevate overall resident well-being. They hold the promise of a thriving future, driven by ecological urbanism and humanisation initiatives that represent a qualitative shift in the urban development paradigm across the Arabian Peninsula. The chapter infers that the development approach surpasses superficial greenery to develop a city that adapts to its environment and maintains natural system connections responsive to its specific context. The Green Riyadh Program, known as the "most ambitious urban greening project in the world," is one of Riyadh's four giga projects launched in 2019 by King Salman Bin Abdulaziz, the Custodian of the Two Holy Mosques, and seeks to position the city amongst the top 100 most liveable cities globally. Encapsulating the significance of these projects in shaping a climate-resilient and equitable city, the chapter concludes by casting light on the deeper meaning of ecology, as propelled from a philosophical and scientific base of knowledge, to drive further research into networks of action in support of the sustainability of the paradigm shift in the Kingdom of Saudi Arabia.

Keywords Sustainable urban transformation · Green Riyadh Program · Vision 2030 · Quality of life · Humanisation · Green infrastructure · Ecology

E. Husukic (✉) · J. Handal · M. Moscatelli
Department of Architecture, Prince Sultan University, Riyadh, Saudi Arabia
e-mail: ehusukic@psu.edu.sa

A. Cheshmehzangi et al. (eds.), *Designing Nature into Cities*, Urban Sustainability, https://doi.org/10.1007/978-981-95-9382-8_10

1 Introduction

The nature in cities has profoundly transformed throughout history. This evolution mirrors broader socio-cultural and technological shifts. The rise of cities brought with it a mental divide between humans and the natural world, altering how we think about our place in the environment. As Ian McHarg argues, the problem of man and nature is not one of providing a decorative background for human play, or even ameliorating the grim city *"it is necessity of sustaining nature as source of life, milieu, teacher, sanctum, challenge, and most of all, unknown in the self, the source of meaning"* (McHarg, 1969, p. 19).

The fragility of the planet and its resources demands speculative design approaches, rather than the conventional solution validation through technical means. Envisioning an alternative urbanism requires development of a new sensibility, one capable of integrating and negotiating the inherent tensions between ecological imperatives and urban development (Mostafavi et al., 2010).

In the context of the Anthropocene, urban sustainability has become both a theoretical imperative and a policy priority. The increasing recognition that cities are at the forefront of climate resilience, biodiversity loss, and human well-being has given rise to a range of theoretical frameworks and large-scale environmental interventions. Although rethinking the city through the lens of nature is not a new phenomenon, as the garden city movement was an urban planning trend in Europe from the 1910s onwards, since the mid-1990s various ideas emerged with implications for how we design and plan cities in the twenty-first century. The current state of globally pressing issues led to more progressive strategies and concepts, with the quest for a looser, emergent urbanism more akin to the real complexities of cities and offering an alternative to the rigid mechanism of centralist planning.

Consequently, landscape urbanism (Waldheim, 2006), ecological urbanism (Hagan, 2015; Mostafavi et al., 2010), landscape ecological urbanism (Steiner, 2011), and biophilic urbanism (Beatley, 2011) emerged as models for an urbanism that attempts to negotiate the social and ecological complexity found in contemporary (mega) cities, their rapid urbanisation, and tenacious infrastructural issues. These models challenge the conventional understanding of cities and nature as independent by promoting a harmonious connection between urban systems and natural ecosystems. They also present an implicit critique of the traditional divide of disciplines and their inability to offer coherent, competent, and convincing explanations of contemporary urban conditions. A contrasting overview of these urbanism models and their relevance to Riyadh is presented in Table 1.

For many across a range of disciplines, landscape has become both the lens through which the contemporary city is represented and medium through which it is constructed. Urban nature holds great potential to address the multifaceted challenges facing contemporary urban environments and to serve as a catalyst for broader sustainability transitions (McCormick et al., 2025). In the context of urban challenges, green spaces have been widely recognised for their multifaceted contributions to city life. Research highlights that they play a critical role in mitigating

Table 1 Comparison of key urbanism models and their relevance to Riyadh (*Source* Authors)

Urbanism model	Key characteristics	Main proponents	Applicability to Riyadh
Landscape urbanism	Views landscape as the foundational element of urbanism; emphasises systems over buildings	Charles Waldheim	Principles are echoed in open space planning and large-scale greening strategies, such as Green Riyadh Program, focusing on interconnected landscapes and spatial continuity
Ecological urbanism	Integrates ecology into urban design; promotes adaptability, resilience, and environmental literacy	Moshen Mostafavi, Susannah Hagan	Manifested in Vision 2030, Green Riyadh Program, and the Saudi Green Initiative, which aim to restore ecological balance in extreme desert environments
Biophilic urbanism	Advocates human–nature connection in cities; enhances well-being and livability through nature integration	Timothy Beatley	Reflected in ecosystem-based approaches within Saudi Vision 2030 and Middle East Green Initiative, integrating ecological functions into infrastructure
Landscape ecological urbanism	Synthesises ecology and design disciplines; views cities as evolving ecosystems	Frederick Steiner	Applied in projects enhancing quality of life through nature-based solutions, especially in Green Riyadh Program and public realm enhancements across the city

climate change, protecting environmental resources, enhancing social cohesion within communities, promoting both physical and mental well-being, and fostering sustainable economic development over the long term. Hagan argues for adaptive processes rather than domination, advocating for "living with" rather than "living over" (Hagan, 2015, p. 26).

It is critically important to increase the quantity of urban green spaces and to improve the quality and accessibility of green spaces, either by establishing new urban green areas or by modifying the characteristics and functions of existing ones. Greening strategies and interventions can be implemented across urban infrastructures and individual buildings at various scales. In addition to protecting the environment and maintaining natural systems, greening infrastructures can also contribute to helping the economy thrive, humanising cities, and improving the quality of life. Therefore, greening infrastructures in high-density cities must accommodate multiple functions and programs to deliver environmental, social, economic, and health benefits. Furthermore, a solid understanding of environmental technologies is crucial for addressing the challenges facing contemporary cities. This is exemplified by the work of a few forward-thinking built environment practices such as Arup and Buro Happold, whose practices reflect a serious commitment to addressing the planet's ecological fragility. They emphasise the vital role of multidisciplinary collaboration, advocating for partnerships between designers and engineers where engineers contribute as equal partners, rather than merely supporting the aesthetic visions of architects. In doing so, we may move beyond further gestural "greening" that fails to meaningfully contribute to the transformation of the city into a sustainable, artificial ecosystem. Going further, it has become increasingly evident that the practice of ecological urbanism necessitates a carefully calibrated integration of top-down ordinance and expertise with bottom-up participation and ideas. Such alliances converge over a shared conceptual and discursive framework to support meaningful collaboration (Hagan, 2015).

Within the overarching frame of governance, a deeper understanding of ecology emerged in the 1970s, moving beyond viewing nature merely for its instrumental or "use" value to humans. This perspective emphasises the intrinsic role of all living beings in sustaining the interconnected web of life (Devall & Sessions, 1985). It is a call for a balanced act between the environmental paradigm and the social paradigm *sin qua non* for a sustainable vision. Central to this approach is the recognition of life's networked nature, principles that should underpin institutions designed to expand, nurture, and sustain ecological and social systems (Capra, 2004, 2016). As Timothy Beatley (2011) points out, the green urban agenda frequently overlooks the essence of "green", neglecting "the life-enhancing and wonder-expanding dimensions of nature itself" (Beatley, 2011, p. 16). These concepts offer a philosophical foundation for reimagining urban governance where individual responsibilities and collective actions converge to foster innovative and lasting ecological solutions.

In response to the rapid rate of transformation, the Kingdom of Saudi Arabia goes beyond conventional methods of planning. A deep cultural memory of building in an extreme climate, combined with a massive state-led development, has generated a planning culture that one could characterise as a bold shift. Thus, large-scale strategies have been employed to effectively respond to emerging conditions that not only take account of the ecological impact of rapid urbanisation but also provide the necessary resources and restorative actions for the well-being and recreation of the citizens. Not surprisingly, this boldness of action in Saudi Arabia, as well as across the Arabian Peninsula, mirrors progressive attitudes towards climate response for an enhanced urban future.

Against the backdrop of Vision 2030, which seeks to foster healthier urban environments by expanding green open spaces, improving walkability, and reducing pollution, this chapter examines the historical development of the city of Riyadh, focusing on the evolving relationship between urban growth and the natural environment. Amongst the key policy directions that complement these goals is the Humanisation Initiative, introduced by former Riyadh Mayor Abdulaziz bin Ayyaf in 2017. This initiative aims to enhance the quality of urban life through a variety of people-centred strategies that seek to activate public space, strengthen social identity, and restore vibrancy to the urban realm by focusing on the human scale of city design. At the intersection of theoretical discourses on emerging models of urbanism and landscape transition, the chapter assembles an empirical inquiry into data sources to elucidate the historical perspective on how a city and its inhabitants are co-evolving with nature, and how all developments and redevelopments are embedded in such a co-evolution. Culminating in an in-depth analysis of contemporary environmental initiatives, and in particular the Green Riyadh Program, this chapter underscores its pivotal role in the broader urban transformation of the city. To this end, the transformative greening initiatives and the Green Riyadh Program are contextualised within global urban greening and sustainability frameworks, the regional environmental agenda, and broader national transformation agenda, particularly Vision 2030 and the Saudi Green Initiative (2021). Given the scale, climatic context, and alignment with broader global, regional, and national agenda, Riyadh has the potential to become a global benchmark for greening arid megacities—a living laboratory for studies in nature-based urban resilience (Habiby & Yaseen, 2025).

2 Riyadh in History: Nature, City, and People

Riyadh, the capital of Saudi Arabia, is situated in the heart of the Arabian Peninsula on a high desert plateau 600 m above sea level. Its growth has been determined by the natural features of the land, with Wadi Hanifa to the west and Wadi al Sulai to the east, both of which continue to influence the way the city is expanding. The name Riyadh originates from the Arabic word *rawdah*, meaning garden or meadow that recalls a time when seasonal wadis like Hanifa and al Sulai nourished fertile palm groves and fields in an otherwise arid landscape (Harrigan, 2016). From a walled settlement of less than 1 square kilometre in 1920, Riyadh has transformed into a sprawling modern capital of 1500 square kilometres (Al-Hathloul, 2017) that now has more than 7 million residents. With a hot desert climate where summer temperatures often exceed 45 °C, this rapid urbanisation has strained the city's natural systems and has shaped and reshaped its relationship with the environment.

The city of Riyadh was founded on the ruins of several structures around 1740 (Al-Hathloul, 2017). Although designated as the capital of the second Saudi state in 1824 by Imam Turki bin Abdullah, Riyadh only gained prominence after Abdulaziz Al-Saud launched his campaign to unify modern Saudi Arabia in 1902 (Facey, 1992). The physical transformation of the city began in the 1930s when King Abdulaziz built a palace and administrative complex, al-Murabba', 2 km north of the city (Al-Hathloul, 2017). This period coincided with a broader phase of sedentarisation, as nomadic populations began settling in and around urban centres like Riyadh (Al-Hathloul & Edadan, 1993). By the early 1950s, the old city was no longer intact, with its walls removed, the governor's palace and main mosque rebuilt, and streets widened for motor vehicles (Facey, 1992).

Historically, residents of Riyadh shared a deep connection with the landscape (Fig. 1). The city was once bordered by green oases, filled with irrigated palm groves, lush fields, orchards, and flora and fauna that were the epitome of an ecosystem that balanced culture and the environment. These green spaces were not merely scenic; they offered comfort, connection, and a sense of peace—places where urban life unfolded. As the city grew, especially after the 1950s, that green belt began to fade. One by one, the gardens and groves were replaced by built-up areas. What was once a vibrant, living edge around the city slowly disappeared, leaving behind a memory of a greener, quieter Riyadh.

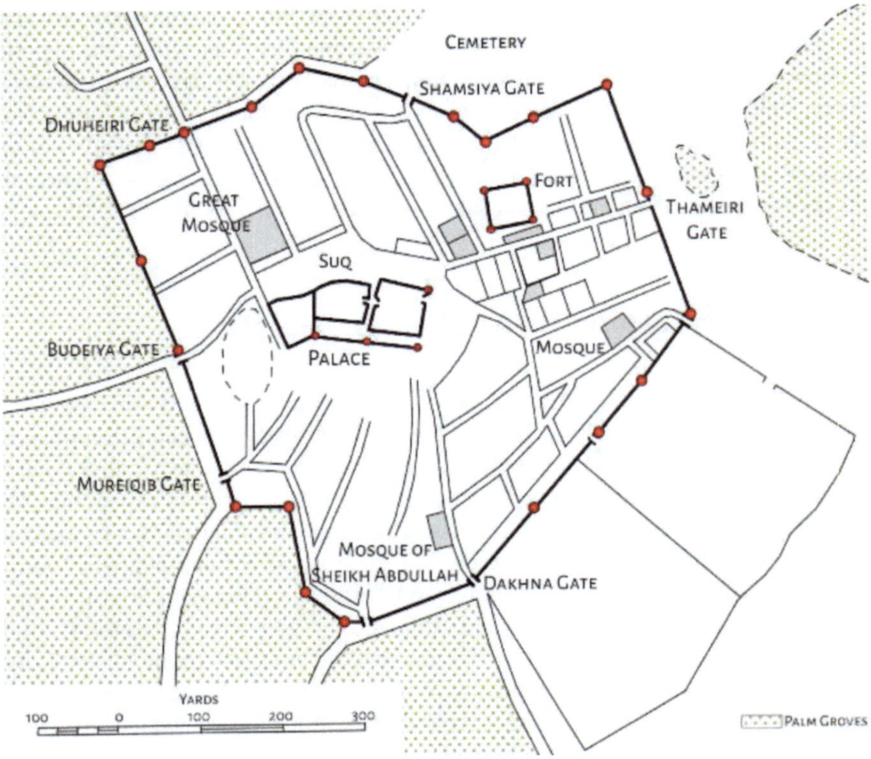

Fig. 1 Map of the walled town of Riyadh from cc. 1922 surrounded by palm groves (*Source* Elaboration by authors)

Between the early 1950s and the late 1960s, a comprehensive development plan initiated by King Saud ushered in a shift towards modern architecture, redefining Riyadh's urban identity. As the population steadily increased during the 1960s, so did the pace of spatial expansion. The authentic desertscape was increasingly replaced by infrastructure, buildings, wide roads, and concrete surfaces that altered both the city's form and its relationship with once integrated nature.

Building on this momentum, the city entered a new phase of urban development in the late 1960s, when the Ministry of Interior commissioned Doxiadis Associates to prepare a master plan for Riyadh. Approved by the Council of Ministers in 1973, the Doxiadis Master Plan consolidated the trends of the 1950s and 1960s, introducing a 2-by-2-km superblock grid system designed to accommodate the city's outward expansion and prioritise vehicular movement (Al-Hathloul, 2017).

Shortly thereafter, the oil boom of the 1970s accelerated these transformations. Riyadh was soon described as the largest construction site in human history. As landscape architect Richard Bodeker observed, *"Oil wealth suddenly catapulted Saudi Arabia, known to the Romans as Arabia deserta, from centuries of being a Bedouin state with an archaic social structure, into the modern world"* (Bodeker,

1997, p. 88). However, in a sensitive desert environment, this rapid growth had significant consequences. Car dependency increased and walkability declined, dehumanising the urban fabric and setting the stage for the sustainability challenges Riyadh faces today. In a country where nature is considerably more sensitive, rapid development brings with it the disadvantage of a progressive expansion of the desert. These patterns of growth, shaped by infrastructure-heavy development and environmental insensitivity to local context in planning, design, and construction, intensified environmental pressures, with long-term implications for both ecological integrity and human well-being.

The period after 1970 is characterised as one of planned national development (Al-Hathloul & Edadan, 1993). This era, led by the establishment of the Arriyadh Development Authority (ADA) in 1984, marked a pivotal moment in shaping the city's development trajectory and building a foundation of sustainable planning mechanisms that were later continued and expanded by the Royal Commission for Riyadh City (RCRC). In the 1970s, the High Commission orchestrated the planning and construction of one of the earliest modern green cities, the Diplomatic Quarter (DQ). Situated to the west of Riyadh, the DQ spans approximately 800 ha and was designed to accommodate around 25,000 residents (Abdullatif et al., 1985, p. 3). This self-contained district is centred around a linear axis that houses the embassies and weaves in five residential neighbourhoods. The design of the Diplomatic Quarter (DQ) integrates both the freedom of the desert landscape and the formality of the Islamic garden. To a great extent, the principles of the Saudi culture steered its design philosophy. With 30% of the area allocated to green space growing outwards towards Wadi Hanifa, the design not only enhances the aesthetic quality of the district but also plays a key role in facilitating natural cooling (Habiby & Yaseen, 2025). Given Riyadh's dependence on desalinated seawater for its water supply, the DQ took a pioneering step in the 1970s by establishing its own greywater treatment facility. This facility recycles water from the embassies to irrigate the district's expansive green infrastructure, reflecting an early commitment to water conservation and sustainable practices (Royal Commission for Riyadh City, 2025). Richard Bodeker, the landscape architect, emphasised that *"sustainable landscape design in dry climates must consider natural resources. This applies to both a balanced water management and the application of design criteria"* (Bodeker, 1997, p. 87).

Although the landscaping for the Diplomatic Quarter in Riyadh and other projects in and around Saudi Arabia's capital have established themselves as good examples of landscaping amidst the arid environment, the rapid pace of urbanisation continued to place significant pressure on the natural landscape. The most substantial growth of Riyadh occurred in the 1980s. While the dominant direction of growth till the 1980s was towards the east and west, in the late 1990s, the city began to expand in all directions. From 1980 to 1990, the city's population grew significantly, reaching 2.3 million inhabitants. As a result, densities have decreased and land allocation per capita has increased, putting pressure on the city's limited natural resources and resulting in an unsustainable growth and development pattern (UN Habitat, 2019).

In contrast to a close-knit, energy-saving, inward-looking traditional urban fabric of Islamic city, staged amongst oases and farms, the development of the 1980s generally took the form of Western induced individual blocks—villas, "walk-ups", or towers—at a density usually much lower than the traditional practice that started in the 1970s. This structure and typology inevitably resulted in large, unshaded roadways that no longer functioned as convection channels but rather as sources of radiant and reflective heat. These adverse effects on the indigenous harmonious open space systems were frequently compounded by the choice, location, and maintenance of plant material. Incidental open spaces were often designed to be too large, exposing vast surfaces of paving to solar radiation, heat build-up, and glare. Meanwhile, the beneficial application of tree planting to improve the climate was often neglecting their value in shading buildings, pedestrians, and cars, increasing humidity and filtering wind-borne dust (Filor, 1988).

As early as 1988, Seamus W. Filor denounced the new Westernised planning, design mechanisms, and imported lifestyle patterns by highlighting the importance of understanding and responding to the local physical, social, and cultural environment when designing in Saudi Arabia. He went on to call for designs of varied character and form that provide contemporary solutions to contemporary problems, not mere stale caricatures or essays in neo-vernacularity—solutions that are responsive to more enduring social and environmental values (Filor, 1988).

By the late 1990s, Riyadh was a very fast-growing city with a population of more than 4 million and an annual growth rate of 8%. In response to the rapid pace of urbanisation, the late 1990s marked a turning point with the launch of MEDSTAR project (Metropolitan Development Strategy for Arriyadh) by ADA—a comprehensive planning framework aimed at guiding the city's sustainable growth through integrated systems, policies, and long-term development strategies (Al-Hathloul, 2017).

The urban fabric that emerged in Riyadh, shaped by planning decisions often transferred wholesale and insufficiently adapted to the region's climatic and cultural conditions, highlights the enduring consequences of disregarding local context. The prevalence of wide roadways, limited shading, and low-density development contributed to increased heat accumulation and eroded the human scale of the city. These early planning misjudgements offer critical lessons for contemporary urban development: the necessity of climate-responsive design rooted in place-specific realities. It is within this context that recent greening initiatives emerge, not only as environmental interventions but as strategic responses to past planning shortcomings and as pathways towards a more sustainable and liveable urban future.

3 Setting Ground for the Large-Scale Greening

The rate of urbanisation has surged during recent decades especially in arid regions including the Middle East where city populations have nearly doubled since the 1960s above the global urbanisation rate of 55% (World Economic Forum, 2024).

The fast-paced growth of cities creates new environmental challenges. The 2019 UN Habitat report highlighted several key challenges facing Riyadh's development. Fragmentation and lack of cohesion within the city structure, socio-ecological disparities, and environmental sustainability topped the list as major issues emanating from the insufficient resilience of the city. Restoring balance to Riyadh's social, ecological, and economic systems aligns seamlessly with the goals of Vision 2030. As a guidance from UN Habitat, *"Riyadh must invest in the creation of new green spaces and in restoring the ecologically sensitive areas to combat natural and man-made environmental threats, as it continues to grow into a metropolis"* (UN Habitat, 2019, p. 92).

A range of ambitious strategies, urban planning policies, and actual interventions have started to be put into action to revive the verdant infrastructure of the past antecedent to Vision 2030 proceedings. The Humanisation Initiative, introduced by former Riyadh Mayor Abdulaziz bin Ayyaf in 2017, aimed to enhance the quality of urban life through a variety of people-centred strategies. *"The concept of 'humanising cities' is about enhancing the human dimension in cities. It transforms a city from being a place that is silent and desolate to a setting that is vibrant and alive. It allows people to construct their own personal narratives and for the community to establish a distinct social identity"* (Bin Ayyaf, 2022, p. 223).

One of the components of this initiative is an environmental development that focuses on a range of projects and programmes designed to restore the ecological balance within the urban fabric of the city. Efforts have included limiting the spread of concrete infrastructure and extensive road networks by expanding green spaces, which not only improve the city's visual appeal but also offer shade and help mitigate extreme heat. Additionally, the initiative emphasises the preservation of natural resources, the conservation of wildlife, and the promotion of overall environmental health (Bin Ayyaf, 2022, p. 231). Due to its importance the concept of humanisation became one of the programs and objectives of the Kingdom's Vision 2030, with its forward-looking approach to enhancing the quality of life in cities.

Wadi Hanifa was one of Riyadh's most significant ecological projects prior to the launch of Vision 2030 (Fig. 2). Through a comprehensive development strategy, over 4000 square kilometres of the Wadi have been reimagined as a key ecological corridor and urban park, offering vital environmental and recreational value to the city. In 2010, the Arriyadh Development Authority received an Aga Khan Award for Architecture for rehabilitation and improvement work on Wadi Hanifah, one of the first of the awards to be given for an environmental project. *"The Wadi Hanifah Wetlands Project eloquently demonstrates an alternative ecological way of urban development. It shows how a major natural phenomenon which, through the course of urbanisation, became a litter-strewn and dangerous place—a scar on the face of the capital city—can be transformed by sensitive planning attentive to social values and imaginative infrastructure-driven landscape solutions"* (Aga Khan Award for Architecture).

These greening projects are not simply environmental or urban interventions but also embody deeply held cultural and spiritual values. The importance of trees, shade, and public space is planted in both Islamic teachings and Bedouin culture,

Fig. 2 View of Wadi Hanifa in 2022 (*Source* Photo taken by the authors)

where hospitality, caretaking of the land, and the common good are paramount principles (Fig. 3). Emphasising public areas as focal points for planting underscores a cultural continuity that merges environmental responsibility with social identity. Furthermore, research on green infrastructure in arid environments (e.g. Al-Ghamdi & Salimi, 2020) supports the idea that integrating local cultural values promotes both ecological resilience and social acceptance of urban greening projects, such as the Green Riyadh Program.

It is clear that Riyadh offers the greatest range of environments that could be enjoyed by its inhabitants. The genius act of incorporating the ecological systems with the urban, social, and cultural landscapes resides in the innovative vitality of the government projects that are driving the green agenda forward.

4 Green Riyadh: From a Distant Dream to a Reality

The Greening of Riyadh is embedded in Saudi Arabia's Vision 2030, a policy framework aimed at economic diversification, environmental regeneration, and quality-of-life improvements, and marks a critical step towards raising Riyadh into the list of the top 100 most liveable cities globally. Three giga environmental initiatives in Riyadh stand out due to their scale and integration within the urban fabric (Fig. 4).

Fig. 3 The parks are designed in a way that provides family compartments—Wadi Hanifa Wetlands, Riyadh, Saudi Arabia (*Source* AKAA/Arriyadh Development Authority)

Taken together, these infrastructural pillars and investments are admirable in their sheer ambition and scale in the environmental regeneration agenda. As one of the pillars, King Salman Park, with over a million square metres of green mixed-use space, is set to become not only the largest urban park in the world but also "the most connected and 'Smartest' urban Park" (Park, K. S., n.d., p. 20). In partnership with King Abdullah University for Science and Technology (KAUST), it is bringing a bold vision to life: a world's first cognitive park (Habiby & Yaseen, 2025). This one-of-a-kind space will feature a comprehensive digital twin—a virtual replica designed to monitor, manage, and enhance the park's functions in real time, blending environmental sustainability with cutting-edge smart technologies. The Sports Boulevard is another ambitious greening initiative which transforms a former utility corridor into a vibrant green thoroughfare. It is a greening of an artery extending 135 km across the city of Riyadh, connecting Hanifah valley in the west with Al Sulai Valley in the east. Wadi Hanifa still holds significant potential to address Riyadh's green space deficit, and the Diriyah Gate Development Authority is currently restoring it as a vital natural corridor and fully integrating it into the city's environmental planning strategies. The Wadi al Sulai to the east of the city is presently only used for stormwater management/drainage systems without providing any recreational benefit to the citizens of Riyadh. As a main node of the Sports' Boulevard project, its future rehabilitation will contribute to restoring the ecological balance of the region.

Green Riyadh Program, the quintessence of the environmental agenda, is one of the largest ambitious urban afforestation projects in the world (Fig. 5). It was

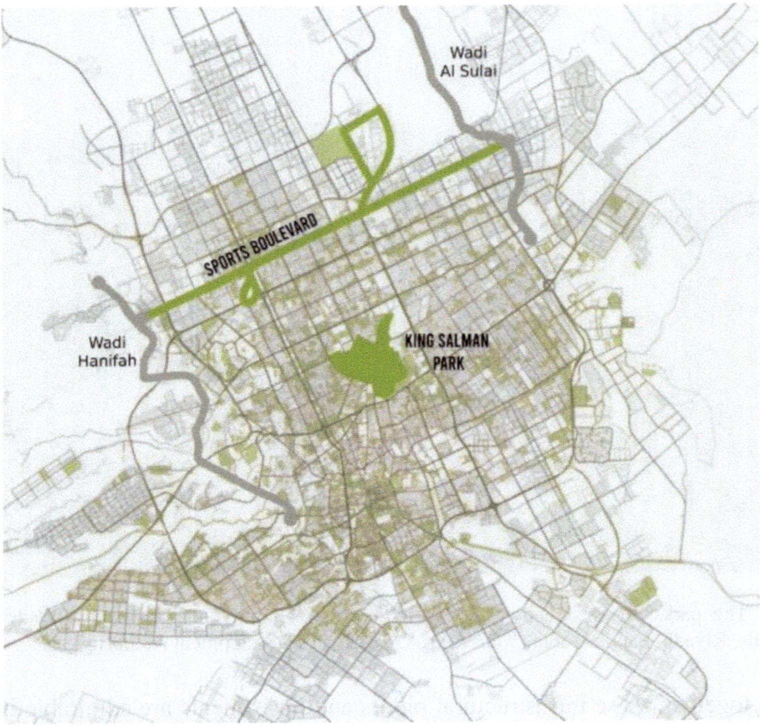

Fig. 4 Riyadh's Green Oasis: King Salman Park and the Sports Boulevard. The map illustrates two major green infrastructure projects—King Salman Park at the city's core, and the Sports Boulevard, a future linear park connecting the western and eastern parts of Riyadh (*Source* Elaboration by the authors)

launched on 19 March 2019, by the Custodian of the Two Holy Mosques King Salman Bin Abdulaziz, at the initiative of His Royal Highness Prince Mohammed Bin Salman Bin Abdulaziz, Crown Prince and Prime Minister. Deeply embedded within Saudi Arabia's broader national transformation agenda, particularly Vision 2030 and the Saudi Green Initiative (2021), it articulates a shift towards environmental responsibility, climate action, and improved quality of life. Moreover, the program aligns with key international frameworks, including the United Nations Sustainable Development Goals (SDGs)—particularly SDG 11 (Sustainable Cities and Communities) and SDG 13 (Climate Action) and the New Urban Agenda developed by UN Habitat.

The initiative aims to plant over 7.5 million trees across Riyadh by 2030 (Fig. 6), increasing the per capita green space from 1.7 to 28 square metres, substantially more than the World Health Organization's recommendation of 9 square metres per inhabitant. It also seeks to expand the total green coverage in the city from 1.5% to 9%, significantly enhancing environmental quality and contributing to a better quality of life for residents. Operating under the supervision of the Royal Commission for Riyadh, Australian consultant EGIS has been appointed to oversee the

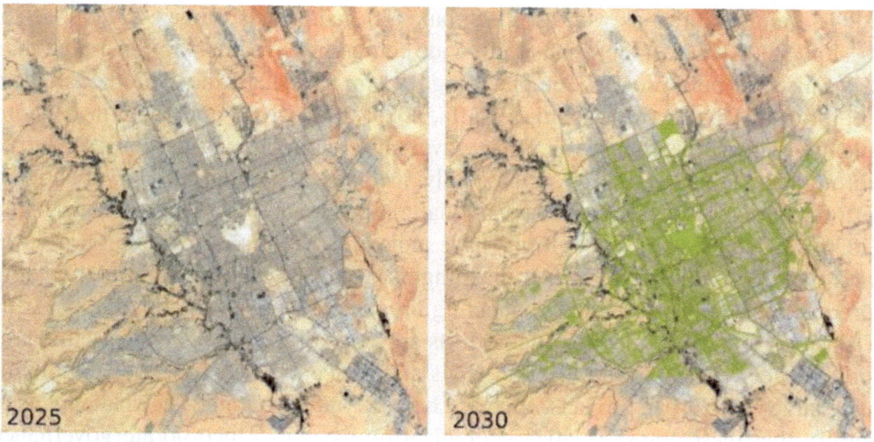

Fig. 5 Green spaces in Riyadh: comparison between the existing green areas and the projected expansion under the Green Riyadh Program (*Source* Elaboration by the authors)

Fig. 6 Road network and urban park greening in Riyadh (*Source* Google Earth, March 2025)

implementation of the Green Riyadh Program. The consultants indicate that, upon completion, the number of trees planted in the greater Riyadh area will match the total planted across all of Europe during the same timeframe (Elsheshtawy, 2022, p. 282). Far from being a typical beautification effort, Green Riyadh Program encompasses city and neighbourhood afforestation projects while using 99 selected tree species that suit the city's environment. Unlike many landscaping projects in the region, which often focus solely on roads and highways, this initiative also targets public spaces such as mosques, schools, healthcare centres, government buildings, and parking areas.

According to Anne Habiby and Lama Yaseen (2025) by developing a citywide network of parks and boulevards, the Green Riyadh Program anchors Riyadh's Greenwave (Habiby & Yaseen, 2025). Under the authority of the Royal Commission for Riyadh City (RCRC), the program encompasses the entire urban area, including all 16 municipalities as well as more than 120 main roads (Green Riyadh, 2025; Vision 2030, 2019). It represents a citywide comprehensive intervention, rather than a localised or pilot program, emphasising the aim to reshape the overall ecological footprint of Riyadh (Moscatelli & Raffa, 2023). This broad approach, in its complexity, reflects the changes in urban governance and planning that have occurred throughout Saudi Arabia as it inevitably involves a multidisciplinary fusion of efforts amongst relevant bodies at national, regional, and local levels. Since Vision 2030 was launched, policy reforms have aimed to decentralise planning processes and promote participatory urban planning at the local level, making the planning process more transparent and connecting urban development to community needs (Alasiri et al., 2025). Coordination efforts are constantly underway to ease integration between these decentralisation goals and the centralised strategies that dominate large-scale urban projects, such as Green Riyadh Program. The challenge is to ensure that these transformative ecological projects also incorporate governance frameworks that facilitate active community involvement in designing and managing urban green spaces over the long term (Alasiri et al., 2025; Ledraa & Aldubikhi, 2025).

Green Riyadh relies entirely on recycled water for its planting efforts, creating a nature-based network that goes beyond aesthetics to play a vital role in enhancing air quality and lowering urban heat levels. It targets a reduction in urban temperatures by 2 °C in the summer and aims to decrease the heat reflected from the earth's surface by 15 °C. Additionally, it seeks to reduce the carbon dioxide levels by 6%, leading to a significant improvement in air quality. This initiative also focuses on biodiversity, establishing ecological corridors to promote bird and other wildlife that are supportive of plant propagation. Economically, through its vast cooling and shading networks, Green Riyadh is expected to reduce electricity consumption by approximately 650 GWh annually by 2030, generating an estimated SAR 71 billion in savings in healthcare costs, reduced electricity usage, enhanced real estate values, and the substitution of potable water with treated wastewater for irrigation. The program also involves establishing regulations and urban policies to encourage afforestation in both public and private developments while actively engaging the community in greening efforts. Socially, the project enhances community well-being by developing open spaces for socialising, sports, and outdoor activities (Habiby & Yaseen, 2025, p. 343; Green Riyadh, 2025).

The Greening initiative is backed by two parallel arrays of enablers, hard and soft ones. Green funds and incentives, research and development, business modelling, regulations and codes, governance systems, communication, and community engagement form the ventures of the soft enablers. While hard enablers subscribe mainly to infrastructural, supply, and operational progress chains summed up in the setting up of nurseries, soil enhancement measures, and a highly elaborate treated sewage effluent network (TSE). Through proactive management, community participation, and innovative technologies, this project exemplifies Saudi Arabia's commitment to environmental sustainability and greening the desert for future generations (Fig. 7).

Fig. 7 Components of the Green Riyadh Program (*Source* Green Riyadh elaborated by authors)

Green Riyadh also carries a symbolic weight. It is an assertion of a post-oil future, where environmental leadership is part of national identity. By transforming a sprawling, car-centric city into a greener, healthier metropolis, the project reflects a paradigm shift in the region's urbanism from consumption towards sustainability. In their overarching impact, Riyadh's greening initiatives promote walkable green connectivity. From an urban planning perspective, they hold considerable potential to reduce reliance on car-based transportation and promote the development of a walkable "15-minute city" (Habiby & Yaseen, 2025). The King Abdulaziz Public Transport Program (Riyadh Metro) that was inaugurated by Royal Decree in 2012 to reshape the city's mobility and opened in 2024 increased connectivity of green spaces within the city. From the start, the Riyadh Metro was conceived as a fundamental sustainability pathway for the city, unlocking new growth while decarbonising transport, a key to the Kingdom's goal to achieve carbon neutrality by 2060. The Riyadh Metro represents a significant evolution in global urban transportation enhancing the city's connectivity (Al Hosain & Alhussaini, 2021). In a regional context, Riyadh's greening efforts are positioned as a flagship initiative within the Middle East Green Initiative, launched in 2021 to foster cross-border environmental cooperation amongst Arab states. This regional framing underscores the geopolitical importance of environmental leadership and stewardship in the Arabian Peninsula.

Within its visionary scope and environmental ambition, the Green Riyadh Program presents some long-term challenges to be addressed within its research goals. On the one hand, the initiative's reliance on 100% recycled wastewater for irrigating 7.5 million trees through a 1350 km high-capacity network represents a novel and efficient response to water scarcity (AIPH, n.d.; Vision 2030, 2019). However, sustaining this system will depend on the resilience of the water infrastructure, especially as urban demands grow. Although treated water is currently sufficient, future pressures could strain the supply without efficient and innovative water treatment solutions. Moreover, the vegetation palette, while determined following profound historical and environmental research into regional flora and fauna systems, may benefit from further research into ecological diversity and potential vulnerability to pests or disease (Elkatoury & Alazba, 2024). Lastly, the maintenance of extensive green areas across all 16 municipalities will require steady funding, strong governance, and avenues for reliance on social responsibility to ensure long-term impact (Moscatelli & Raffa, 2023; Ledraa & Aldubikhi, 2025).

Transformation from desert to green city revives all the thwarted hopes of those in the search for a more beneficent, healthy, and safe environment. Saudi Arabia and Riyadh await a large-scale demonstration of a desertscape developed with wisdom, skill, and taste: a step towards Green dream. No more propitious circumstances could exist than those in the Green Riyadh Program today. The challenge is here and now.

5 Discussion and Implications

From the small fortified desert town of less than half a million inhabitants to the metropolis of the twenty-first century characterised by hyper modernism, Riyadh has drastically changed its relation with nature. Historically, the urban footprint of the city was a sustainable ecosystem, tightly intertwined with its existing ecological structures, such as the wadi, parks, and agricultural groves. This aligns with Ian McHarg's (1969) perspective that tradition founded applied ecology as a base for function and aesthetics in landscape. However, those features, which created the unique identity of the city, have been lost over time. Increased urban density has amplified the impacts of climate change and biodiversity loss. The urban expansion that menaces the open green spaces today is a typical national problem. As in the cases of other cities worldwide, if growth is uncontrolled, sporadic nature will slowly recede to be replaced by growing islands of development. These will in time coalesce into a mass of low-grade urban tissue, having eliminated all-natural beauty and diminished rare excellence, both historic and modern.

Undoubtedly, the three environmental mega-projects of Riyadh (King Salman Park; the Sports Boulevard; and Green Riyadh) will bring about green and open spaces creating a multipurpose green network. Aside from their aesthetic value, a "green canvas" has the potential to generate various benefits for people including urban biodiversity, improved air quality, additional green spaces, and in consequence, an enhanced quality of life. The vision of Riyadh to become a contemporary oasis draws not only from its history rooted in the wadi system that has sustained the region for centuries but also from a deeper cultural and spiritual symbolism. This aspiration reflects Islamic values and Arab heritage, serving as a reminder of a time when people lived in harmony with nature. It embodies both teachings of the Holy Quran and traditional Bedouin principles of environmental respect. Today, this legacy shapes Riyadh's ambition to evolve into a sustainable city (Harrigan, 2016).

Amongst the global greening efforts, Riyadh's Greening initiatives stand out due to their scale, climatic context, and integration with a broader national and regional transformation agenda. Globally, cities such as Singapore, Copenhagen, and Melbourne have positioned green infrastructure at the core of their urban sustainability agendas. Singapore's "City in a Garden" model integrates natural systems into the built environment, employing vertical greenery and urban biodiversity strategies (Newman et al., 2009). Similarly, China's "Sponge Cities" initiative emphasises nature-based solutions to manage stormwater and urban heat (Yu, 2025). Riyadh's environmental giga projects align with these global trends, yet its desert context introduces unique challenges particularly water scarcity and extreme temperatures that are driving locally adapted greening solutions. Unlike ad hoc or decentralised green interventions, Green Riyadh is embedded in Saudi Arabia's Vision 2030, a policy framework aimed at economic diversification, environmental regeneration, and quality-of-life improvements, and marks a critical step towards raising Riyadh into the list of the top 100 most liveable cities globally. It is part of both the Saudi Green Initiative and Middle East Green Initiative.

Parallelism with global initiatives drives avenues for benchmarking and fostering institutional learning across capacity building, cross-sectoral governance, and long-term sustainability potential. Effective inter-agency coordination, strategic partnerships, and robust monitoring systems are critical mechanisms employed within the Vision 2030 agenda towards sustaining momentum beyond initial implementation in support of adaptive management (Annual Report Vision 2030, 2024). Avoiding reliance on top-down models and its adverse impact on detachment from local needs have been addressed by Green Riyadh Program through an integration of governance and community engagement in its set of components (Fig. 7). Further research into pathways for participatory approaches and grassroots involvement is central to the goals of Green Riyadh. Riyadh's greening initiatives thus offer a distinctive contribution to global urban sustainability, combining ambitious scale, cultural heritage, and institutional integration. Balancing top-down coordination with inclusive community engagement can position Riyadh as a replicable model for arid megacities worldwide, promoting resilient and vibrant urban futures.

6 Ways Forward?

This brief historical repertoire of Riyadh outlined the interrelated relationship between urban growth and the natural environment. It depicted an image of an environmentally thriving future, driven by the ecological initiatives of the ambitious Vision 2030 with its paradigm shift towards the question of sustainability and quality of life. The ecological greening movement of the Kingdom of Saudi Arabia is enabling the evolution of processes that can produce a noble life-enhancing physical environment and an ennobling model lending exemplary lessons yet to be drawn on.

The essence of this conclusive note is to catalyse profound inquiry, aimed at advancing the intricacies of urban governance processes in support of the Kingdom's commitment to a sustainable environmental paradigm. As evidenced in the aforementioned sections, the underlying concepts coherently elucidate the merits of establishing governance systems drawing on ecological paradigms as the nexus for converging individual responsibilities and collective actions towards fostering innovative and sustainable ecological solutions. Within the evident collective action of government initiatives, as controllers to the process and the product, and in a shift towards a balanced network system, what role could social organisations play to define their central responsibility towards a better quality of life and a sustainable urban environment? Could a Business Improvement District be a prototype of collective networks between the controllers and the producers of the built environment and its systemic ecological sustainability?

In acquiring an ecoliteracy mindset, what actions could individuals undertake to mainstream the greening network of Riyadh and mobilise it from a germination phase within parks, valleys, green corridors, and institutions to a propagation phase into every peripheral street and every household—inherent in the very evolutionary characteristic of nature? How networks of Volunteer Systems can be weaved across institutions to organise the recruitment, training, systematic engagement, and retention of volunteers? In support of ecological initiatives, what networks of monetary and non-monetary incentives can be extended in the field of urban policies and urban management in relevance to the ecological paradigm? And at a national level, what material and immaterial infrastructural assets can be built for a holistic shift in both social and environmental paradigms?

… Not an end … but rather an inquisitive start, widening and deepening an identification with ecology…exhortations to embracing the networks of nature as the driving forces for sustainability!

Acknowledgements We would like to express our sincere gratitude to Dr. Zahir A. Othman, architect and urban planner, for his invaluable guidance and support throughout the development of this chapter. The authors would also like to acknowledge the support provided by Prince Sultan University.

References

Abdullatif, A., Othman, Z., & Ward, G. (1985). The Diplomatic Quarter and Ministry of Foreign Affairs Staff Housing Project, Riyadh. In M. Bentley Sevcenko (Ed.), *Designing in Islamic Cultures 5: Large housing projects* (pp. 3–11). Aga Khan Program for Islamic Architecture.

AIPH. (n.d.). *Riyadh, Saudi Arabia: Treated irrigation water network. A green city case study*. Retrieved July 27, 2025, from https://aiph.org/green-city-case-studies/riyadh-saudi-arabia-water/

Alasiri, F., Dąbrowski, M., Rocco, R., & Forgaci, C. (2025). The impact of recent policies on the transformation of local participatory urban planning in Saudi Arabia. *Urban Science, 9*(3), 69. https://doi.org/10.3390/urbansci9030069

Al-Ghamdi, S. G., & Salimi, M. (2020). Climate change impacts on critical urban infrastructure and urban resiliency strategies for the Middle East. *Sustainable Cities and Society, 54*, 101948. https://doi.org/10.1016/j.scs.2019.101948

Al-Hathloul, S. (2017). Riyadh development plans in the past fifty years (1967-2016). *Current Urban Studies, 05*(01), 97–120. https://doi.org/10.4236/cus.2017.51007

Al-Hathloul, S., & Edadan, N. (1993). Evolution of settlement pattern in Saudi Arabia. *Habitat International, 17*(4), 31–46. https://doi.org/10.1016/0197-3975(93)90027-A

Al Hosain, N., & Alhussaini, A. (2021). *Evaluating access to Riyadh's planned public transport system using geospatial analysis*. King Abdullah Petroleum Studies and Research Center. https://doi.org/10.30573/KS--2021-DP10.

Beatley, T. (2011). *Biophilic cities: Integrating nature into urban design and planning*. Island Press.

Bodeker, R. (1997). Gardens in the desert: A landscape architect in Saudi Arabia. In W. O'Reilly (Ed.), *Sustainable landscape design in arid climates: Proceedings of a symposium jointly organised by the Aga Khan Trust for Culture [and others]: Held at Dumbarton Oaks on 7 December 1996* (pp. 87–95). Aga Khan Trust for Culture.

Bin Ayyaf, A. B. M. (2022). *Local governance and the municipal sector: Challenges and missed opportunities. Riyadh as a model*. Tarah International.

Capra, F. (2004). *The hidden connections*. Knopf Doubleday Publishing Group.

Capra, F. (2016). *The systems view of life*. Cambridge University Press.

Devall, B., & Sessions, G. (1985). *Deep ecology-living as if nature mattered*. Gibbs Smith, Publisher.

Elkatoury, A., & Alazba, A. A. (2024). Irrigation water demand management-based innovative strategy: Model application on the Green Riyadh initiative, Saudi Arabia. *Water, 16*(24), 3559. https://doi.org/10.3390/w16243559

Elsheshtawy, Y. (2022). *Riyadh: Transforming a desert city*. Routledge.

Facey, W. (1992). *Riyadh, the old city: From its origins until the 1950s*. Immel.

Filor, S. W. (1988). Landscape architecture in Saudi Arabia. *Landscape Research, 13*(2), 23–28. https://doi.org/10.1080/01426398808706259

Green Riyadh. (2025). Retrieved May 8, 2025, from https://riyadhgreen.sa/

Habiby, A., & Yaseen, L. (2025). Riyadh's urban greenwave: Fostering city resilience through large-scale greening. In A. Arora et al. (Eds.), *Climate-resilient cities: Priorities for the Gulf Cooperation Council countries* (pp. 341–352). Springer Nature Switzerland.

Hagan, S. (2015). *Ecological urbanism: The nature of the city*. Routledge.

Harrigan, P. (2016). *Riyadh: Oasis of heritage and vision*. Medina Publishing.

Ledraa, T., & Aldubikhi, S. A. (2025). Unveiling the value of green amenities: A mixed-methods analysis of urban greenspace impact on residential property prices across Riyadh neighborhoods. *Buildings, 15*(12), 2088. https://doi.org/10.3390/buildings15122088

McCormick, K., Kiss, B., Palgan, Y. V., Bulkeley, H., Davis, M., Raven, R., Luque-Ayala, A., & Hörschelmann, K. (2025). *Urban nature: New directions for city futures*. Cambridge University Press.

McHarg, I. L. (1969). *Design with nature*. In *Published for the American Museum of Natural History [by] the Natural History Press*.

Moscatelli, M., & Raffa, A. (2023). Green infrastructure in arid urban contexts. Transitioning ecologies beyond Green Riyadh. *AGATHÓN | International Journal of Architecture, Art and Design, 13*, 75–86. https://doi.org/10.19229/2464-9309/1362023

Mostafavi, M., Doherty, G., & Harvard University (Eds.). (2010). *Ecological urbanism*. Lars Müller.

Newman, P., Beatley, T., & Boyer, H. (2009). *Resilient cities: Responding to peak oil and climate change*. Island Press.

Park, K. S. (n.d.). *THE DESTINATION FOR INVESTMENT*. https://kingsalmanpark.sa/-/media/files/ksp-investment-brochure-11.pdf

Royal Commission for Riyadh City (RCRC). (2025). *RCRC Programs and Projects*. Diplomatic Quarter Development Program. Retrieved May 8, 2025, from https://www.rcrc.gov.sa/en/projects/diplomatic-quarter-development-program/

Steiner, F. (2011). Landscape ecological urbanism: Origins and trajectories. *Landscape and Urban Planning, 100*(4), 333–337. https://doi.org/10.1016/j.landurbplan.2011.01.020

UN Habitat. (2019). *Riyadh city profile*. Ministry of Municipal and Rural Affairs and United Nations Human Settlements Programme. https://unhabitat.org/sites/default/files/2020/03/riyadh.pdf

Vision 2030. (2019). *Green Riyadh*. Retrieved July 27, 2025, from https://www.vision2030.gov.sa/en/explore/projects/green-riyadh

Vision 2030. (2024). *Annual Report*. Retrieved August 8, 2025, from https://www.vision2030.gov.sa/media/25042025ev1/En-Annual%20Report-Vision2030-2024.pdf

Waldheim, C. (Ed.). (2006). *The landscape urbanism reader*. Princeton University Press.

World Economic Forum. (2024, June 5). *Urban transformations. Cultivating nature, cooling cities: An urban future grown to last*. Retrieved May 1, 2025, from https://www.weforum.org/stories/2024/06/nature-cool-cities-sustainable-urban-future/

Yu, K. (2025). Sponge Planet: Nature-based Infrastructure for Climate Adaptation Beyond Concrete. *Landscape Architecture Frontiers, 13*(5). https://doi.org/10.15302/J-LAF-0-020052.

Reimagining Urban Futures: Lessons for Designing Nature into Cities

Sara Alidoust, Wendy Y. Chen, Richard Fuller, and Ali Cheshmehzangi

Abstract *Designing Nature into Cities* advances an urgent agenda for reimagining the human–nature relationship in urban planning and design through the lens of environmental ethics and social justice. The book examines how cities have historically engaged with nature, the current pressures on this connection, and innovative approaches for building sustainable urban futures. It is a call to action for the planning, design, and development of places that humans live, work, and play across the world.

Structured into three interconnected parts, the volume articulates actionable pathways—embedding biophilic metrics in master plans, advancing participatory planning, integrating data for evidence-based decision-making, and securing policy coherence and cross-sectoral and multidisciplinary coordination and collaboration—that move beyond symbolic greening towards sustainable, context-sensitive practice. Drawing on case studies from across the world, including examples from both developed and developing countries, the book demonstrates how nature-based solutions can be embedded across scales: from building materials and planting strategies to public space, heritage precincts, and metropolitan areas. The part "*Edge*

S. Alidoust (✉)
School of Architecture, Design and Planning, University of Queensland, Brisbane, QLD, Australia
e-mail: s.alidoust@uq.edu.au

W. Y. Chen
Department of Geography, University of Hong Kong, Hong Kong, Hong Kong

R. Fuller
School of the Environment, University of Queensland, Brisbane, QLD, Australia

Centre for Biodiversity and Conservation Science, The University of Queensland, Brisbane, Australia

A. Cheshmehzangi
School of Architecture, Design and Planning, University of Queensland, Brisbane, QLD, Australia

School of Architecture and Urban Planning, Hunan University, Changsha, China

College of Architecture and Urban Planning (CAUP), Tongji University, Shanghai, China

A. Cheshmehzangi et al. (eds.), *Designing Nature into Cities*, Urban Sustainability, https://doi.org/10.1007/978-981-95-9382-8_11

Ecologies: Coasts and Deserts" particularly highlights the growing role of nature-based solutions in strengthening climate resilience in vulnerable regions.

This book serves as a valuable resource for students and researchers in geography, environmental studies, architecture, design, and urban planning. It offers both conceptual clarity and practical guidance to transform urban vulnerabilities into opportunities. Through aligning ecological imperatives with social equity, the book informs strategies for creating urban systems that are resilient, inclusive, and culturally responsive.

Keywords Nature-based design · Nature-based urbanism · Nature-based solution · Urban ecological design · Nature-based governance · Urban sustainability · Biophilic design · Placemaking · Climate resilience · Participatory planning · Urban heritage · Coastal resilience · Desert urbanism · Human–nature relationship · Blue carbon ecosystems

1 Designing Nature into Cities: A Reflection

This book is the third of our trilogy focused on "Nature in Cities", an idea that we could continue to add more volumes and worldwide case studies. The first volume focuses on Nature in Cities, Nurturing Cities (Cheshmehzangi et al., 2026a), the second introduces the concept of Nature-Based Urbanism (Cheshmehzangi et al., 2026b), and this third volume focuses on Designing Nature into Cities. Nature within cities has undergone profound transformations in the history of cities and urban life. This evolution reflects broader socio-cultural and technological shifts in urban history. Urban development introduced a mental divide between humans and nature and reshaped the perception of human relationship with the environment (Benton-Short & Short, 2013). This mindset has recently started to shift, at least within academic discourse. Nature-based solutions are no longer viewed as an "optional luxury" in urban environments (Montgomery, 2013, p.120). They are increasingly recognised as essential for addressing key urban challenges such as climate change and driving urban regeneration. However, integrating nature into cities cannot occur in isolation. Growing evidence highlights the need for co-creation and collaboration in cities, aiming at balancing ecosystem and social benefits. This calls for integrated approaches to design, planning, and implementation that embed nature through participatory and collaborative processes (Frantzeskaki, 2019).

Designing nature into cities is explored in this edited volume, which is structured into three interconnected parts. Part I, *Design Methods and Material Practices*, underscores the integration of nature as a core design principle. It advocates for incorporating plants across multiple scales, from individual buildings to urban landscapes, emphasising the ecological and cultural significance of native species in supporting biodiversity and reinforcing local identity. Their inclusion strengthens design effectiveness and resilience. This section informs a new framework for embedding living elements into architectural and urban planning practices as tools for advancing urban sustainability, particularly in subtropical cities.

A critical theme emerging here is the material paradox in nature-based placemaking. While many placemaking initiatives worldwide are publicly promoted as eco-conscious and nature-driven, there is often a disconnect between symbolic sustainability and ecological realities. This underscores the need to recognise the complexities that shape sustainability in placemaking practices. Designing nature into cities requires understanding sustainability as a negotiated, assemblage-based material process—rather than a purely visual or discursive outcome. Materials are not inherently sustainable or unsustainable and should not be treated as passive components within a fixed framework. Their role and impact are dynamic, shifting according to treatment methods, environmental conditions, cultural values, and governance structures.

Part II, *Heritage, Identity, and Urban Transformation*, investigates the intersections of three critical themes: placemaking, nature-based initiatives, and urban heritage and identity. Understanding these synergies is vital for developing equitable strategies that redefine human–nature relationships and promote just, resilient, and ecologically balanced urban futures. Urban heritage plays a pivotal role in shaping identity, spatial structure, and public culture. In many historic cities worldwide, nature represents both a lasting heritage and a living future. It offers valuable insights to policymakers seeking to strengthen local identity, enhance community well-being, and support meaningful human–nature connections.

Drawing on case studies from diverse contexts—from Europe to the Middle East—this part explores how nature-based approaches can enrich urban design while activating placemaking strategies to strengthen community engagement with heritage and historic fabric. These interventions illustrate how neglected and undervalued heritage areas can be reactivated and integrated into broader sustainability agendas.

Finally, Part III, *Edge Ecologies: Coasts and Deserts*, addresses the growing importance of nature-based solutions in building climate resilience within vulnerable regions across the world. Drawing on the analysis of a wide array of strategies, policy frameworks, and global case studies, this section explores potential nature-based interventions across contrasting environments—from coastal shorelines to desert cities. It outlines pathways for leveraging blue carbon ecosystems as multidimensional governance tools for resilience, with particular emphasis on the sustainable transformation of coastal tourist cities under climate change pressures. Another central theme of this part of the book is the examination of transformative greening initiatives, particularly the Green Riyadh Program. The discussion moves beyond superficial greenery, advocating development approaches that enable cities to adapt to their ecological context and strengthen connections with natural systems.

2 What We Have Learned: Key Insights and Takeaways

A central insight emerging from this edited volume is the importance of integrating native species into building design and urban development to strengthen local biodiversity and identity. This approach must be supported by policy guidelines and industry standards that incorporate site-specific ecological knowledge early in the design and planning processes. Such integration ensures long-term sustainability and place-based responsiveness, particularly where climatic conditions demand adaptive and flexible strategies.

Design and planning practices must also move beyond superficial eco-aesthetics towards what this book terms “slow sustainability”. This involves prioritising ecological repair, co-production of knowledge, and reinforcement of local material ecosystems. Addressing material politics is equally essential and goes beyond the question of what materials are used to how they operate within broader networks of power. These considerations underscore the socio-political dimensions of sustainability and highlight the need for more equitable and context-specific approaches.

Researchers must critically examine prevailing design methodologies and compare them with alternative approaches. As highlighted in this volume, there is an urgent need to shift from viewing design merely as a tool for taming nature within the urban fabric towards embracing design actions that engage with the inherent complexity and unpredictability of the natural world in its many forms (Pollastri et al., 2021). Envisioning alternative design methodologies requires being capable of negotiating tensions between ecological imperatives and urban development. This involves moving beyond deterministic planning towards adaptive, responsive frameworks that integrate ecological resilience with cultural and social priorities.

At a broader scale, strategic planning must integrate technical and participatory approaches. This means embedding biophilic metrics into city master plans while employing bottom-up practices for adaptive design and creating roadmaps for tailoring nature-based solutions to local contexts. Participatory planning must be institutionalised to ensure inclusive and equitable outcomes for all community groups. This requires building grassroots capacity, establishing inclusive decision-making platforms, and ensuring that benefits (and burdens) associated with initiatives are shared equitably.

Systems thinking and integrated planning and decision-making processes are essential for mainstreaming these approaches. Policy coherence and cross-sectoral coordination are fundamental. Aligning efforts across government agencies, NGOs, donor institutions, and local communities will create enabling environments for nature-based solutions to thrive (Dushkova & Haase, 2020). For example, the role of blue carbon ecosystems in addressing urban climate vulnerability should not be viewed solely as a tool for carbon offsetting; rather, it must be understood as a comprehensive resilience mechanism operating across scales, systems, and governance structures. Their restoration and protection often involve multiple sectors, including water authorities, environmental agencies, the tourism industry, and landowners. Establishing clear legal status, participatory governance, management authority, and financial mechanisms can transform blue carbon systems into formal governance instruments. Beyond ecological restoration, blue carbon serves as a medium for advancing social justice and ecological governance and links climate resilience with equity.

Finally, research institutions play a pivotal role in developing evidence-based monitoring systems to evaluate ecological, social, and economic outcomes. Moving forward, investment in localised pilot projects, developed through collaboration between research and practice sectors, is essential. These can serve as living laboratories for testing nature-based solutions under varying climatic conditions and generate insights that inform adaptive management and future interventions.

3 Concluding Remarks

Designing Nature into Cities reconnects ecological processes with social and political dynamics in cities. It highlights the transformative role of planning and design in reinforcing the interconnectedness of all life, human and non-human, on our planet. Cities are inherently complex, yet planning has historically sought to reduce this complexity, sometimes with adverse effects. A critical step is to acknowledge the depth of urban complexity through systems thinking approaches instead of erasing it (Marshall, 2011).

This book reaffirms that sustainability is inseparable from social justice and epistemic diversity. The findings reveal that power structures, manifested through formal institutions with unsustainable visions and top-down development plans, often undermine sustainability. Such approaches not only degrade natural ecosystems but also produce adverse social impacts, entrenching cycles of inequality and poverty. These dynamics highlight the urgent need for participatory, justice-oriented planning, where decisions are informed by environmental, social, and cultural priorities.

The future of placemaking depends not only on the materials we use, but on understanding how people and materials co-constitute places and how these relationships evolve over time. Current practices often reveal deeper structural tensions in how sustainability is imagined, institutionalised, and implemented.

There is a need to redefine the human–nature relationship in city making discourse based on environmental ethics, with nature-based solutions offering a promising pathway towards this goal. While challenges remain, global best practices demonstrate that collaboration among research institutions, community stakeholders, and policy frameworks can transform vulnerabilities into opportunities towards creating more resilient, sustainable, and equitable systems.

References

Benton-Short, L., & Short, J. R. (2013). *Cities and nature*. Routledge.

Cheshmehzangi, A., Alidoust, S., Chen, W. Y., & Fuller, R. (Eds.). (2026a). *Nature in cities, nurturing cities*. Springer.

Cheshmehzangi, A., Alidoust, S., Chen, W. Y., & Fuller, R. (Eds.). (2026b). *Nature-based urbanism*. Springer.

Dushkova, D., & Haase, D. (2020). Not simply green: Nature-based solutions as a concept and practical approach for sustainability studies and planning agendas in cities. *Land, 9*(1), 19.

Frantzeskaki, N. (2019). Seven lessons for planning nature-based solutions in cities. *Environmental Science & Policy, 93*, 101–111.

Marshall, S. (2011). Planning, design and the complexity of cities. In *Complexity theories of cities have come of age: An overview with implications to urban planning and design* (pp. 191–205). Springer Berlin Heidelberg.

Montgomery, C. (2013). *Happy city: Transforming our lives through urban design*. Penguin UK.

Pollastri, S., Griffiths, R., Dunn, N., Cureton, P., Boyko, C., Blaney, A., & De Bezenac, E. (2021, June). More-than-human future cities: From the design of nature to designing for and through nature. In *Proceedings of the 5th Media Architecture Biennale Conference* (pp. 23–30).

MIX
Papier aus verantwortungsvollen Quellen
Paper from responsible sources
FSC® C105338

If you have any concerns about our products,
you can contact us on
ProductSafety@springernature.com

In case Publisher is established outside the EU,
the EU authorized representative is:
Springer Nature Customer Service Center GmbH
Europaplatz 3, 69115 Heidelberg, Germany

Printed by Libri Plureos GmbH
in Hamburg, Germany